It Hurts Down There

It Hurts Down There

*The Bodily Imaginaries
of Female Genital Pain*

CHRISTINE LABUSKI

Published by State University of New York Press, Albany

© 2015 State University of New York

All rights reserved

Printed in the United States of America

No part of this book may be used or reproduced in any manner whatsoever without written permission. No part of this book may be stored in a retrieval system or transmitted in any form or by any means including electronic, electrostatic, magnetic tape, mechanical, photocopying, recording, or otherwise without the prior permission in writing of the publisher.

For information, contact State University of New York Press, Albany, NY
www.sunypress.edu

Production, Diane Ganeles
Marketing, Kate R. Seburyamo

Library of Congress Cataloging-in-Publication Data

Labuski, Christine, author.
 It hurts down there : the bodily imaginaries of female genital pain / Christine Labuski.
   p. ; cm.
 Includes bibliographical references and index.
 ISBN 978-1-4384-5885-4 (hardcover : alk. paper)
 ISBN 978-1-4384-5886-1 (paperback : alk. paper) — ISBN 978-1-4384-5887-8 (e-book)
 I. Title. II. Title: Bodily imaginaries of female genital pain.
 [DNLM: 1. Vulva—United States. 2. Vulvodynia—psychology—United States. 3. Shame—United States. 4. Social Stigma—United States. 5. Vulvodynia—ethnology—United States. 6. Women's Health—United States. WP 200]
 RG261
 618.1'6—dc23
                                              2015001460

10 9 8 7 6 5 4 3 2 1

CONTENTS

List of Figures	vii
Acknowledgments	ix
Prologue: A Fourteen-Foot-Tall Vagina	xiii
1. Insinuation: A Biocultural Condition	1
2. Examination: Clinical Interpretations of Vulvar Pain	35
3. Accumulation: The Materiality of Absence	77
4. Manifestation: (Un)conscious Presencing	119
5. Integration: Coming Together or Falling Apart	161
6. Generation: Novel Morphologies	195
7. Evaluation: Concluding Thoughts	235
Epilogue: Collaboration	251
Notes	259
References	275
Index	311

FIGURES

Figure 1. Vulvar pain nomenclature   37

Figure 2. A clinical vulva   39

Figure 3. The cotton swab, or "Q-tip," diagnostic test   40

Figure 4. A modified vestibulectomy   45

Figure 5. A full set of vaginal dilators   72

Figure 6. Muscles of the pelvic floor   102

ACKNOWLEDGMENTS

This book has benefited from the wisdom, insights, and bodily experiences of far too many people to mention by name—from the women on whose bodies I first learned to do pelvic exams, and from whom I learned to talk about sex as a health care professional, to the members of my dissertation committee, who grew increasingly comfortable uttering the word "vulva" as we discussed my work. But without the following people, places, and institutions, *It Hurts Down There* would have remained a discussion between a far smaller group of people than the circulation of a book makes possible.

Not a word of this book could have been written had nearly one hundred women with vulvar pain refused my queries, my curiosity, and my presence in their lives. You are brave and magnificent, and I am grateful beyond measure for the imaginaries you shared with me. I hope that each of you recognizes at least some small part of your experience in the pages that follow and that this book might help you to share that experience with a slightly wider world. I am also indebted to the physicians, physical therapists, nurses, and medical assistants with whom I worked for thirteen months. Their cooperation with and support for my research and the space they made for my nonafflicted body in their exam rooms, surgical suites, and treatment sessions provided this project with a richer set of layers than I had dared believe was possible.

My time in the vulvar clinic was first made possible through a Sexuality Research Fellowship, administered by the Social Science Research Council, and I am forever indebted to Diane di Mauro for that early and pivotal support. My research was also supported by the University of Texas at Austin, in the forms of a Liberal Arts Graduate Research Fellowship; a David Bruton, Jr., Graduate School Fellowship; a University Co-op/ George H. Mitchell Award; and a Women's and Gender Studies Dissertation Fellowship. I am grateful, too, for the enthusiastic welcome and support I received from the Center for the Study of Women, Gender, and Sexuality at Rice University, where this project thrived under the benevolent direction of Rosemary Hennessy. Other compatriots in Houston from whom my work benefited include Samar Farah Fitzgerald, Melissa Forbis, Colton Keo-Meier, Emma Kate Lyders, Brian Riedel, and Angela Wren Wall. At the University of Arkansas, where the manuscript gestated while I taught (and taught and taught), I want to thank Rob Brubaker, Lisa Corrigan, Kirstin Erickson, Dave Fredrick, Jennifer Hoyer, Kelly O'Callaghan, Lindsay Puente, Laurent Sacharoff, Raja Swamy, and Ted Swedenburg for their friendship and generous support of my ideas. Finally, this manuscript might never have made it to press had Virginia Tech's Program in Women's and Gender Studies and Department of Sociology not given me the time, support, and financial assistance that it needed in its final stages. I especially want to thank Aaron Ansell, Rachelle Brunn, Toni Calasanti, Tom Ewing, Laura Gilmann, Anthony Kwame Harrison, Bernice Hausman, Sharon Johnson, Ann Kilkelly, Minjeong Kim, Neal King, Sarah Ovink, Anthony Peguero, Katy Powell, Petra Rivera-Rideau, John Ryan, Manisha Sharma, and Barbara Ellen Smith for their encouragement and collegial support. I also owe an enormous debt of gratitude to Saadia Rais, Meagan South, and Rayanne Streeter for their important contributions to the manuscript.

At the University of Texas at Austin, Kamran Ali and Pauline Turner Strong worked together to give this project its initial shape; I am enormously grateful to them for making me a clearer and more critical writer and analyst. I also want to thank Deborah Kapchan for introducing me to critical body studies, as well as Laura Lein and John Hartigan for their contributions to the dissertation. Austin colleagues who made their indelible marks on these pages include writing group comrades Melissa Biggs, Claudia Campeanu, Alice Chu, Liz Lilliott, Jessica Montalvo, Apen Ruiz, and Guha Shankar, as well as early cheerleaders Alisa Perkins and Lisa Schergen. Can Aciksoz, Sergio Acosta, Jenny Carlson, Beth Bruinsma-Chang, Dan Gilman, Elizabeth Hawthorn-Leflore, Jennifer Karson-Engum, Ritu Khanduri, Shanti Kulkarni, Ken Macleish, Mathangi Krishnamurthy, Brandt Peterson, Rachael Pomerantz, Mubbashir Rizvi, Leela Tanikella, Teresa Velasquez, Halide Velioglu, Mark Westmoreland, and Casey Williamson round out an extraordinary group of colleagues who enhanced my every minute in the Department of Anthropology.

At State University of New York Press, I want to thank Beth Bouloukos, Rafael Chaiken, and Diane Ganeles for seeing the manuscript through to completion; I also want to thank two anonymous reviewers whose careful readings and commentary shaped a much stronger version of the manuscript. Claire Insel and Clair James are deeply appreciated for their expert and timely editorial ministrations.

I have two kinds of family to thank, the one with whom I share a name and the one that has sustained me in my long years away from them. The Labuskis are an amazing bunch, and I am so grateful for all they do for me, and for each other, in the wake of my physical absence. Though anthropologist has been a somewhat murkier career path than the one in nursing that preceded it, they have kept pace with and helped me to celebrate my erratic successes. Other family, who have housed, fed, clothed, nourished, listened to, and basically propped me up any time it was in order, include the Biggs-Coupal household, Genevieve Buentello, Lora and Matt Brown, Andrea and John Kelso, Lori Nickels, Brittany Phillips, John Toole, John Van Voorhies, and Erin Von Feldt. I cannot imagine my life without the delightfully steady hand and companionship of Kathleen McCarthy, who has literally seen me through it *all*. And for everything in between, and everything that spills over the edges; for making me laugh and making me think; for caring about and caring for; for compromise and mutual respect; and for more comfort and joy than I could ever have asked, I want to thank Nicholas Copeland. I am forever lucky that you have my back.

PROLOGUE

A Fourteen-Foot-Tall Vagina

In June 2014, a young American man, visiting the campus of Germany's Tubingen University, got himself wedged inside a stone sculpture in the shape of a vulva. The piece, titled *Pi-Chacán* (roughly, "making love" in Quechua) and commissioned by the university from artist Fernando de la Jara for $173,000, stands fourteen feet tall and has flanked the campus's Institute of Microbiology since 2001. Sensing a rare photo opportunity, the young tourist decided to climb inside the vulva's opening, never doubting his ability to exit the sculpture once his friend had captured his image. The idea lost considerable charm for both young men, however, when it became necessary to enlist emergency personnel—twenty-two firefighters working for several hours—in order to extricate the student from *Pi-Chacán*.

Unsurprisingly, this genital tale immediately went viral (they say you can't make this stuff up); also not surprising, given that most of my social network is aware of my research interests, my inbox and Facebook wall were quickly populated with several versions of the story, allowing me to quickly skim—and later peruse—both the reporting and the comment threads that proliferated steadily throughout the day. Most of the news accounts were matter of fact: the articles I read were succinctly reported, and all contained a basic version of the events that had transpired. And though the headlines ranged from titillating ("Giant Vagina Sculpture Traps US Student in Germany")[1] to vaguely esoteric ("Thoughts, Freud? Vagina Sculpture Traps US Student")[2] to downright cheeky ("It's a Boy! Student Rescued from Vagina Sculpture in Germany"),[3] none were as salacious as the comment threads, where the student was referred to as both a "douche" and a "cunt," where the situation itself was described in "fishy" terms, and where the massive size of the sculpture's orifice was compared to the proliferating "whorishness" of contemporary women. But what became most notable—and least surprising—for me, as I followed the story through its social media death, was the regular and reiterated disconnect between the *vulvar* sculpture to which the majority of articles explicitly referred and the *vaginas* via which every headline trumpeted the story.

Like this story, and the sculpture around which it developed, this book is about vulvas. More specifically, this book is about painful vulvas, vulvas that can no longer be ignored by the bodies that include them or by the people with whom they are in relationships. More broadly, it is about all of the other vulvas, including the one at Tubingen University, that are culturally invisible. Reading the prologue, you might be wondering about my use of the word "invisible" and how it pertains to a massive stone vulva capable of "trapping" an American tourist and gaining internet notoriety in a matter of hours. Though my answer to this hypothetical objection will take shape throughout the book, I introduce the tale of this unfortunate young tourist to draw your immediate attention to the linguistic disconnect of the *Pi-Chacán* story: of how the word "vagina" was recruited to erase and disappear the sculpted vulva. What that practice has to do with vulvar pain conditions, as well as what they both have to do with the female bodily imaginaries at stake in this book, is something about which I hope you will learn a great deal by reading the following chapters. As internal canals and "potential" spaces,[4] vaginas are difficult to render materially visible or graspable, and sculpting an anatomically "accurate" one would demand a distinct perspective and perhaps even medium. True, we might allow that because the student went inside the sculpture, he entered a vaginal space. But a reading of this story that accords value to the entirety of female genitalia, as does this book, insists that this space was *not* the sculpture to which news headlines gleefully referred. The sculpture, as was visually plain to everyone who read about *Pi-Chacán* and the hapless student, was of a vulva.

It Hurts Down There is a book about why vulvas—even fourteen-foot-tall ones—don't make headlines; it is also a book about what it means for all of us when they disappear. Told through the very real, though erratically apparent, condition of female genital pain, this book invites readers to consider both the absence and the hyperpresence of female genitalia and to ask questions about how women's lives are altered by a pain that few of us will name. Feminist author Rebecca Solnit (2014) has suggested that words can be used to either bury or excavate meaning: "If you lack words for a phenomenon, an emotion, [or] a situation, you can't talk about it, which means that you can't come together to address it, let alone change it." This book constitutes a refusal to bury the vulva. Derived via feminist and anthropological inquiry, and carried out for the reasons conveyed by Solnit, my excavation of the vulva strives to provide all of us with the knowledge, will, and tactical skills not only to tell better stories about female genitalia but also to allow them to speak for themselves.

A NOTE ABOUT TERMS

This book is about several types of clinical disease conditions that I collectively refer to as "vulvar pain/disease," "genital pain/disease," *vulvodynia* and *vulvar vestibulitis syndrome* (*VVS*), *vestibulodynia*, or simply "symptoms." Unless otherwise noted, these terms are used interchangeably. I also employ the phrases "genital dis-ease" and "vulvar dis-ease" to refer to a less physiological (and more cultural) condition characterized by the disparagement, awkwardness, and silence through which female genitalia are often experienced in the contemporary United States. All phrases and terms are more thoroughly defined throughout the text, particularly in chapters 1 and 2.

In an effort to collapse the distinction between these two types of "syndromes," I also use the capitalized phrase "Vulvar Disease," which I understand to be a complex interaction between cultural cues and physiological anomalies that accumulate and eventually manifest as intractable and embarrassing genital pain.

Ethnographic accounts of medical conditions routinely employ clinical terms with which readers may not be familiar, and this book is no exception. I have attempted to define all of the terms upon which an understanding of my theoretical argument depends, and it is my hope that interested readers will learn more about those that continue to intrigue them. Also in the tradition of cultural anthropology, *It Hurts Down There* uses a number of concepts and terms that, although derived from the pain conditions under examination, are less clinical and more analytical in nature. In my discussions with readers of this book, I have encountered varying levels of comprehension about some of these terms, ranging from consternation to interdisciplinary curiosity. It is with these generous colleagues in mind that I preface this text by preparing the reader for a variety of terms with which I hope you will remain patient. Some, such as "shutting down," are categories native to the project, meaning that they were uttered and formulated by my informants; quotation marks are used to indicate these terms. Others, such as "dis-ease" and "clean space," are my analytic efforts to describe what I experienced during my time in the field; these terms appear without quotations marks and are—whenever possible—defined on their first use. Still others—including disavowal, obligation, accommodation, and hexis—are concepts borrowed from other theorists that I have adapted for the analysis of vulvar pain conditions. Any misunderstandings on the reader's part are the result of my own inability to translate these concepts adequately rather than the responsibility of my informants or the other theorists whose work has influenced this project.

Finally, the text contains numerous transcripts from interviews I conducted with my informants as well as from the clinical sessions I was invited to observe and document. Many of these transcripts contain unfinished words (indicated by an em dash), verbal hesitations, stammers, pauses (indicated by ellipses), and other indications of discomfort that I believe are integral to an understanding of the awkwardness and embarrassment through which female genital pain is lived. I have therefore left these transcripts as intact as possible.

1

INSINUATION

A Biocultural Condition

———•———

> The worst name anybody can be called.
> —Germaine Greer, *The Female Eunuch*

In the spring of 2003, the *Journal of the American Medical Women's Association* published a study entitled "A Population-Based Assessment of Chronic Unexplained Vulvar Pain: Have We Underestimated the Prevalence of Vulvodynia?" At seven pages, the article concisely reports the findings of a telephone survey of over three thousand Boston-area women who were asked about symptoms that the authors defined as "chronic vulvar pain":

   a) burning in the genital area for 3 months or longer with or without chronic itching,
   b) knifelike or sharp pain in the genital area for 3 months or longer with or without burning or itching, or
   c) excessive pain on contact when inserting tampons, during sexual intercourse, or during pelvic examinations that lasted for 3 months or longer. (Harlow and Stewart 2003, 83)

The survey was part of a larger project conducted by the Harvard School of Public Health, nestled between pilot research demonstrating an 18 percent prevalence rate for vulvar pain and a clinic-based follow-up study intended to correlate women's reported symptoms with objective evidence of disease. The authors of the study (an epidemiologist and a researcher-physician) reported that "the pathophysiology of these conditions [and] . . . the magnitude of this problem . . . [were] largely unknown" (82), but that "the true incidence of generalized and localized vulvar dysesthesia" could not be determined "without a complete medical history and physical examination to rule out *other causes* of genital discomfort" (83; my emphasis).

For these researchers and their clinician audience, "other causes of genital discomfort" include vulvar dermatoses, malignancies, inflammatory conditions, postoperative or postinjury neurological complications, and recalcitrant or atypical presentations of yeast or bacterial infections of the vagina and vulva. Although etiologically and pathologically varied, what links these causes of genital discomfort is the medical certainty that they are physiological and that a resolution of the pain can and should be achieved through pharmacological or surgical means. Reasons for distinguishing between chronic vulvar pain and its possible look-alikes are, in this framework, related not to the nature of its source (i.e., physiological) but rather to the project of constituting and delineating a distinct category of genital disease. Noting that this "highly prevalent condition . . . is associated with substantial disability" (87), the authors conclude with the hope that suitable prevention strategies can be gleaned from a better understanding of chronic vulvar pain's distinct "etiological pathways" (87).

This book investigates the contemporary landscape of vulvar pain in order to illuminate a distinctly faceted set of causes and prevention strategies. Chronic and unexplained vulvar pain is indeed an increasingly legitimate medical condition. Though exact causative mechanisms remain unclear, researchers have postulated a wide range of possible factors—including neurological injury, genetic susceptibility, and altered hormone expression—and have developed an expanding array of variably effective treatment options (Bachmann et al. 2006; Danby and Margesson 2010; Haefner et al. 2005; Leclair et al. 2007; NIH 2012). This book argues that vulvar pain is also, and perhaps primarily, a bodily *experience*, mired in discourses of pollution and taboo that severely restrict a woman's ability to accurately describe her symptoms. In the words of one informant, vulvas are virtually "off-limits," including to the women whose bodies they distinguish. Women with vulvar pain bow to the weight of censoring social discourses as they struggle with language, postpone and avoid clinical consultations, and refuse treatment options that necessitate physical encounters with their genitalia. Rooted in a "highly prevalent" social genital dis-ease, these behaviors index a "disability" no less "substantial" (Harlow and Stewart 2003, 8) than the painful conditions described by Harvard's researchers.

In this book, I treat vulvar pain as a condition that is simultaneously clinical and cultural, an approach reflecting my training in both feminist anthropology and gynecological medicine.[1] Medically, it is an affliction severe enough to preclude vaginal penetration, sitting down for longer than a few hours, and wearing pants or other forms of fitted clothing. Because the etiology of genital pain remains unclear, the diagnostic experience is a protracted one: clinicians attempt to identify the source and best course of treatment for symptoms they don't fully understand, while affected women's lives are irrevocably altered (Ventolini 2013). Vulvar pain is a cultural condition in that most women in the contemporary United States[2]—including symptomatic ones—lack an idiom or sufficient vocabulary through which to name and describe the parts of their body that are in pain. Frequently severe, this limitation is coassembled with a host of cultural institutions and practices through which frank descriptions of female genitalia are contaminated, erased, or otherwise muted.

I argue that distaste toward female genitalia is socially conditioned—neither natural nor inevitable; in Butler's words (1993), it can be regarded as a "regulatory schema that produce[s] intelligible morphological possibilities" (14). In the chapters that follow, I explore the cultural and clinical conditions through which this dis-ease condition is achieved, how it "produces and vanquishes bodies that matter" (14); I also elaborate some of the risks with which it is associated and describe strategies through which alternative female genital bodies might be recuperated. In this introduction, I insinuate the contours of female genital dis-ease by positioning it within the symbiotic fields (Bourdieu 1993) of phallocentric and biomedical heteronormativity—discourses and practices that prioritize and assume the routine and penile penetration of female genitalia. Following this, I discuss the importance of feminist analyses that foreground the *intra*genital dynamics of the female sexual body—that is, the important and often hierarchical differences between vulvas and vaginas—and caution that failing to make these distinctions risks missing fundamental insights regarding how female bodies are culturally imagined. Furthermore, I argue that labial anatomies, thought by many to figure only marginally in heteronormative sexual scripts, have the capacity to profoundly disrupt and reconfigure them.

In the final section of this chapter, I offer three concepts without which I do not believe vulvar pain conditions can be properly analyzed: *genital dis-ease*, *unwanted genital experience*, and *genital alienation*. Unpacked in turn, these concepts demonstrate what medical anthropologists often label *bioculture* (Wiley and Allen 2009) and what "new materialists" refer to as *entanglement* (Barad 2007; Coole and Frost 2010): that events and ideas thought to be exclusively social are registered *and* reflected by material bodies, and that disease conditions are constituted by this multidirectional process. In the case of vulvar pain, the paradoxical narratives of excess and inconsequence shape a bodily imaginary in which genitalia are a problem to be avoided rather than confronted. Feminist and anthropological attention, I argue, can complement the clinical strategies of vulvar experts, reorienting symptomatic women away from the former and toward the latter.

FIXING THE PROBLEM

> The way he is, I have to fix the problem.
> —Sharon, Vulvar Health Clinic patient

Sharon[3] uttered this statement to me as we talked over coffee one afternoon, engaged in a formal interview about her vulvar pain. Though symptomatic for over five years, Sharon had been neither properly diagnosed nor adequately treated until the month prior to our interview, when she had finally secured an appointment with the Vulvar Health Clinic (VHC), where, in 2004–2005, I conducted ethnographic fieldwork in order to better understand how symptomatic women and their providers negotiated the difficult "cultural" encounters that talking about genitalia required. For thirteen months, I observed and transcribed patient visits at the clinic and then followed these women through other aspects of their treatment regimens, including surgery and physical therapy sessions. This was in addition to formal interviews and informal conversations designed to glean as much information as possible about the wider contexts of vulvar pain. Sharon was emblematic of many of the women I came to know that year: having just consulted with an expert, she was newly optimistic that her symptoms would eventually resolve. But her buoyancy lay over several years' worth of sexual and bodily despair, indexed most poignantly by her report of having "shut down," both emotionally and sexually, a long time ago.

The "he" in Sharon's remark was her husband, and her observation that the problem was hers to fix speaks to one of vulvar pain's most onerous realities. The inability to engage in penetrative intercourse, one of the more notable hallmarks of this disease, engenders a unique set of stressors for patients' intimate relationships. Almost always heterosexually identified, most women adapt to their symptoms by closing off connections to their genital bodies (sometimes for years at a time), participating in painful sex, or engaging in sexual activities that one informant explained were "more about him." I argue that these limited coping strategies are elements of a pernicious and heteronormative social structure from which many straight-identified women must labor to discern their erotic sensibilities.

Immersed in the notion that vaginal-penile penetrative coitus constitutes "real sex" (Kaler 2006), women and couples affected by vulvar pain enact a set of seemingly natural implications: rather than exploring the nonpenetrative or nonvaginal dimensions of their sexual bodies and desire, they defer an active or generative sex life until symptoms can be wholly resolved (Labuski 2014). Sharon's sense that it was *she* who needed to "fix [her] problem"—in other words, be able to "have sex" again—indexes the limited range of options through which many couples live their heterosexuality. In short, her husband's phallocentric desire is the rule, rendering women who cannot comply with routine intercourse the exception. For both members of this couple, Sharon's pain functions as little more than an obstacle in that, rather than providing a motive to learn about what (other) parts of their bodies they might *also* enjoy, it drives them further away from physical intimacy. As I elaborate in subsequent chapters, neither partner is able to perceive the coexistence of vulvar pain *and* a heteronormative sexual relationship. Rather, both believe that they must choose what Sharon's body can tolerate: *either* the pain *or* her husband.

The situation in which Sharon and her husband are embroiled creates a space for what some scholars would call the queering of their heteronormality. Arguing that radical sexuality need not be solely the province of so-called sexual minorities, Beasley (2010) calls for a "transgressive heterosexuality" that "reject[s] . . . simplistic accounts of sexual modes" by "refus[ing] to inculcate socio-political determinism" (208). Stevi Jackson (2006) furthers this analysis with a discussion of meaning, concluding that while such determinisms may govern the intelligibility of (sexual) norms, meaning "is also negotiated in, and emergent from, the mundane social interaction through which each of us makes sense of our own and others' gendered and sexual lives" (112). Here, a "'natural attitude'" (113)[4] toward penetrative intercourse can be called into question through beliefs and behaviors in which couples like Sharon and her husband might choose to engage, thus widening the parameters of what it means to be (hetero)sexually "normal."

For Cacchioni (2007), such practices constitute "sexual lifestyle changes":

> Rather than working towards mastering, strategically mimicking, or carefully avoiding sexual practises, sexual lifestyle changes . . . involve challenging normative definitions of sex and even the overall importance of sexual activity. [They] might involve . . . privileging sexual

activities typically deemed as foreplay, and/or valuing non-goal-oriented masturbation as an acceptable sexual activity on par with intercourse with another person. They also might entail questioning the overall importance placed on sexual relationships, institutions and practices. (310)

In this analysis, pain can become a catalyst for two kinds of shifts, the first being in one's sexual repertoire, toward a set of more comfortable behaviors, and the second in the broader gender dynamics that structure a couple's relationship. As Kaler (2006) argues, vulvar pain produces a category of "unreal women" (50) whose gender identity is threatened by being unable to engage in "an action which makes people into heterogendered men and women" (58). Likewise, Kempner (2014) has shown that this brand of gender transgression extends to the condition of migraine. Women whose debilitating headaches make it hard to care for their children, maintain a household, or participate in sexual activity are pathologized for failing to comply with these gendered norms, particularly by pharmaceutical companies that market remedies that enable them to return to the roles of wife and mother. I want to underscore that both types of pain confound women's gendered identities by thwarting the aims of patriarchal masculinity; the conditions to be cured are defined as much by the (gendered) work in which symptomatic women cannot fully engage—vaginal intercourse and child care—as by the pain itself. Because of this conflation, and as my ethnography makes clear, sexual relationships that are (re)defined in terms of what *her vulva needs* rather than what *his penis expects* can facilitate the reconfiguration of gender norms in other areas of a heterosexual couple's life.

And though these interpersonal tensions take their toll on vulvar pain patients, the struggle in which most of these women are even more intimately engaged is the one with their own bodies. It is these bodies—female,[5] genital, sexual, and in pain—that are the protagonists of this book. I believe that the *intra*personal efforts of women afflicted with vulvar pain should be of concern to feminists for two important reasons: first, because their recuperative efforts locate them on the cutting edge of a radical and nonphallocentric sexuality; and, second, because they provide a uniquely embodied perspective on how female genitalia are lived in the contemporary United States. As anatomy and physiology—as skin, muscles, blood, and nerve endings—the genitalia of women with vulvar pain are "bodies that matter" (Butler 1993) apart from the erotic behavior in which they might or might not engage. But as I demonstrate throughout this book, this singular bodily fact is routinely undermined, in ways both astonishing and ordinary, by the erasure and muting of female genitalia.

MY BIRTH IN THE CLINIC

It's tricky to physically inhabit a part of the body from which you have been otherwise taught to disassociate, through, for example, the "shaming words and dirty jokes" to which Gloria Steinem refers in her introduction to *The Vagina Monologues* (Ensler 2001, xi). Throughout this book, I use the term "dis-ease" to convey the awkwardness of encounters with the vulva, as well as to underscore the role this awkwardness plays in not only our general understanding of vulvar pain but also in its rates of complication and severity. It is the mutuality of these dimensions to which I refer when I use the phrase "Vulvar Disease," which I capitalize in order to provide increased analytical weight to this relationality. As a formal theoretical concept, Vulvar Disease emphasizes the biocultural nature of a physiological pain saturated with gendered meanings and expectations, a real disease shot through with beliefs and sensibilities that contour its progression.

Freud's notorious assertion that "the sight of female genitals gives . . . rise to 'horror, contempt, or pity'" (Gatens 1996, 34)[6] speaks to only one of the more explicit legacies through which women encounter dis-eased genitalia. These affects of disgust are compounded by what Harriet Lerner describes as the "persistent misuse" and substitution of the word "vagina" for "vulva," a practice that "impair[s] a girl's capacity to develop an accurate and differentiated 'map' of her . . . genitals" (2005, 28). When symptoms arise in this unmentionable (and therefore unspecified) place, the familiar act of uttering the words necessary for a focused medical history ("It hurts when I breathe"; "The itching seems to be much worse at night") requires a delicate and difficult set of negotiations among the woman, bodily ignorance, propriety, and the urgency of her painful situation. "Having accurate language to distinguish the vulva from the vagina is crucial for every girl," continues Lerner. "Inaccurate labeling . . . increases shame and complicates healing" (2005, 28; see also Frueh 2003).

It was through my work as a gynecological clinician—working first as a nurse and then a nurse practitioner in Planned Parenthood and other so-called sexual health clinics—that I initially came to speculate about the vulnerability of genitalia from which my patients seemed to be detached. As a health care provider, I was initially concerned about the disease-related outcomes of this detachment: the malignant progression of an undetected vulvar lesion, for example, or the potentially life-threatening complications of a sexually transmitted infection (STI). My fellow clinicians and I routinely lamented that our patients couldn't *talk* about the very same sexual bodies that they physically shared with their partners (Braun and Kitzinger 2001; Devault 1990), and I began to wonder about the wide gaps that existed between what I taught my patients to do and what they later (at times sheepishly) told me they actually did. And though I am not suggesting that these problems are not shared by other areas of medicine, I am saying that genital health matters occupy a distinct cultural sphere, and that both clinicians and patients are challenged to invest in a bodily realm from which the rest of their worlds are often—and actively—*dis*invested.

What I couldn't see, however, during my clinician days, was that I had been taught to assume far too much about the bodily integrity of my patients. A conventional program of college nursing, combined with an emerging feminist consciousness (a decidedly second-wave one[7]), had convinced me that patients simply needed more education to make "healthier" decisions (Metzl and Kirkland 2010), and that the cultural context of their lives—though interesting—was ancillary to their medical situations. In many ways, *It Hurts Down There* is the handbook that I would offer to that young and eager nurse, whose politico-professional stance was chronically at odds with the clinical realities before her. Though she still has plenty to learn, I want to tell her that her instincts were right—that there *is* something amiss and that her patients often can't (or won't) use contraception effectively because they are unable to confront their sexual and genital bodies. I also want her to know that despite her ability to effectively intervene, at least at times, the discomfort and alienation shaping her patients' unwanted sexual situations were far more insidious than her individual efforts could address.

While I was conducting fieldwork, my friend's high school–age daughter called one afternoon to tell me about an academic conference she had just attended at a local university, the focus of which had been sexual assault and domestic violence. Though she had loved the conference, she found herself troubled by a "feeling" she had never before been aware of, one that surfaced as she listened to stories and feminist analyses of these two painful social realities. "It's in my stomach somewhere," she told me. "I don't know; it's this *feeling*." In response, I tried to share with her my own version of that *feeling* and how it had emerged for me during those early and trying years as a nurse. We spent some time commiserating about the other feelings it generated: anger, disgust, helplessness, inspiration (to intervene), vulnerability, and a grim and abiding acceptance of what it (sometimes) means to be a female body in the contemporary United States. As a clinician, I had never been able to properly address this feeling, given that my time with patients typically ranged between five and fifteen minutes. And though I chose to channel it into increasingly creative levels of prescriptive and supportive advice, I continued to notice that my patients' relative (in)abilities to be at home in their bodies almost always ended at their genitalia.

Undaunted by—and eager to account for—this collective reticence, I refocused my efforts toward feminist anthropology, so that I might investigate this genital reluctance through a less individualistic frame. My questions about bodily integrity, when posed in collective terms, lead me to further questions about the cultural sources of our bodily understandings and about the corporeal "maps" through which we do or don't connect with various parts of our anatomy. From this perspective, missing vulvas became both neuropsychological events as well as material instantiations of female sexual inferiority. I see these two dimensions as mutually obliged and, in the following chapters, I analyze them as both separate and interactive phenomena. In widening my anthropological lens to include the physiological functioning of the bodies in question, I not only return to some of my medical roots, I also offer a more complete rendering of how the vulva is made both present and absent through cultural disavowal.

My analysis of vulvar pain is positioned squarely within a feminist politics and in the service of a critical anthropology of the body (Karkazis 2008; Lock and Kaufert 1998; Manderson 2011; Martin 1987; Wentzell 2013). The discursive and material disavowals of female genitalia are structured and routinely sustained by the cultural institutions of patriarchy, heterosexuality, and gynecological medicine. Long disciplined and disparaged (Braun and Tiefer 2010; Muscio 1998), the vulva, I argue, is in need of recuperation at all three levels of bodily experience: individual, social, and political (Scheper-Hughes and Lock 1987). With this book, I begin that project, offering an attention to this genital flesh—in all its vulnerability, alienation, and inconsequence—through which new modes of identification might be possible. In these chapters, I create a space for the vulva to exist for and as itself—as an anatomical, neurological, erotic, vascular, and functional element of a body—and, in this way, contribute to the longstanding feminist project of reimagining female sexuality on its own terms (Braun and Tiefer 2010; Irigaray 1985a, 1985b; New View 2000).

GENITALITY

Sorry your vagina looks like a grenade went off at a deli counter.
—Text of a circulating e-card/meme[8]

Based on the largest existing collection of ethnographic and qualitative data regarding vulvar pain, this book constitutes an important layer of recognition for a condition about which most people know very little. A feminist engagement with vulvar pain is important not only for the patients and clinicians whose stories resonate with those of my research informants but also for sex and gender scholars who share my interest in the critical study of genitalia. As the students in my Sexual Medicine course have discovered, even defining the term "genitals," particularly after one gains an awareness of the variability through which people and bodies live the term, can be a protracted affair. Are uteruses genitalia? Cervixes? Prostate glands? If my relationship with these organs is sexually recreational rather than procreative, are they still "reproductive" organs? What about pubic hair? Do people have more and less genitalia if and when they alter them? (In the first iteration of this exercise, we settled on the definition "what's between our legs," though many of us continued to harbor reservations.)

Many of the questions that vulvar pain raises—regarding sex, gender, and genitalia—are being actively investigated by scholars of transgender identity and intersexuality (Fausto-Sterling 2000; Kessler 1998; Reis 2012; Stryker and Whittle 2006; Valentine 2007). Katrina Karkazis (2008), for example, in an ethnography of surgeons and parents who confront infants with various intersex conditions, demonstrates that both groups recruit genitalia in order to shore up the reality of binary sex. Surgically altering these genitalia, the *Fixing Sex* of her title, enacts a certainty about not only the mutually exclusive nature of male and female bodies but also about the permanence of this distinction. Karkazis convincingly demonstrates that sex is literally constructed via surgical instruments and procedures, and that gendered assumptions about the capacity for penile intromission and vaginal receptivity undergird medical decision making. The narratives of women with vulvar pain complement and extend these arguments by evincing the iterative and evolving nature of the gender-genitals-sex triad. In other words, the genitalia that surgeons presume to have fixed will remain so only if they continue to function; genitalia that are not "usable," in the words of one of my informants, illustrate the ongoing role of gendered praxis in shaping sex. "Real women" (Kaler 2006,

50) and men, in other words, have genitalia that interact with one another in procreative and heteronormative ways.

Though feminists have long paid attention to the female genital body as a site of cultural discourse (Bordo 1993; Irigaray 1985b; Moore and Clarke 1995), a sustained analysis of genitalia as both social *and* biological entities is lacking in the social sciences. Biological vulvas "matter" (Butler 1993) in that they make plain some of the important differences between women, differences that can be elided through other forms of collective organizing (Carrillo Rowe 2008). Women whose vulvas have eroded or been excised, for example, due to disease conditions like those examined in this book, understand their genitalia through body images distinct from those of women whose vulvas have been acquired or enhanced, either through gender affirmation treatment[9] or cosmetic alteration. Though the degree to which female genitalia have been socially disciplined has been well documented, as has the importance of representing and affirming genital diversity (New View 2000), biological variations among women remain undertheorized.

Though based on the experiences of a relatively narrow spectrum of female bodies, this book—a theory of the intragenital dynamics of the female (sexual) body—represents a first step toward such an analysis. A more thorough investigation into the relationships among the vagina, clitoris, labia, perineum, and vulvar vestibule in a feminist sexual politics and practice can frame useful questions such as why women overwhelmingly claim their vaginas over their vulvas, and how these external and internal aspects of female genitalia have become so easily conflated. What, in short, are the implications of a vaginal rather than a vulvar politics? I offer several hypotheses throughout this text, but a great deal of work remains to be done regarding these identification practices, particularly in the context of cosmetic vulvar surgery. Vulvar (self-)censorship is perniciously embedded in popular culture, and it is unclear why feminists have left this practice largely unexamined. My own reaction to the relative dearth of vulvar scholarship has never been one of rebuke. But I remain puzzled by the fact that for every feminist who matter-of-factly reminds us that "it should go without saying that the vagina is not the vulva" (Frueh

2003, 138), there is an equal if not greater number of vulvar "refusals," well characterized by this recent—and particularly exasperated—post on a prominent feminist blog: "I don't care about your stupid vulva, it's all vagina to me" (West 2012).

Additionally, a more thorough engagement with female genitalia, including their biology, allows for more careful analyses of how genital bodies intersect with sexual ones. Though this relationship has rarely been posited as one of neat correspondence, the contexts of trans, intersex, diseased, asexual, and surgically altered genital bodies should render it even less so. The state of one's genitalia does not constitute the bulk of an individual's erotic identity: genitals can be sick, ignored, acquired, aesthetically pleasing, or even absent in ways that cannot always be reduced to a person's "sexual" self. Similarly, erotic sensibilities are not confined to the genitalia and are often distributed across and among a wide array of affects, objects, people, and anatomical locations; as my research informants demonstrate, vulvar pain patients can learn to extend their genital imaginaries to other bodily locations. The degree to which our body maps—or schemata—are biologically inherited remains an open question, but the fact that they are malleable, dynamic, and influenced by experience has now been well established (Berlucchi and Aglioti 1997; De Preester and Tsakiris 2009l; Knoblich et al. 2006). Reducing genitalia, therefore, to their reproductive, sexual, or even functional dimensions can blunt both the meaning and associated affects that might otherwise accompany genital anatomy. Shaped from birth, when "what's between our legs" determines which of two extant gender categories will structure the majority of our lived experience, our personal genital imaginaries are as rich and varied as they are impoverished, owing to a wide range of individual and collective experiences through which vulvas, vaginas, and penises are culturally available.

CRITICAL HETEROSEXUALITY

This book contributes to the literature in critical heterosexuality studies by providing ethnographic evidence of a coital imperative—that is, the narratives through which heterosexual couples prioritize penetration in their sexual repertoires. This evidence, gathered not only through medical consultations and formal interviews but also through intensely personal physical therapy sessions that often included patients' partners, unpacks heterosexual practice in novel ways. By elaborating how the practice of heterosexuality is enabled by both bodily function and compliance, as well as by routine gynecology, I denaturalize the institution itself. Moreover, the stories of my informants provide alternative routes through which male-female sexual relations can be reconfigured, including sexual imaginaries that foreground the vulva.

In the following chapters, I analyze the sexually discursive work done by women with vulvar pain; I use my ethnographic data to suggest that symptomatic women's "interpretation" (Jackson 2006, 113) of routine heterosexuality contains notable amounts of ambivalence. Unable to participate in routine penetrative intercourse, my informants demonstrated a range of problem-solving behaviors, most of which were performed in slow, cautious, and erratic fashion: refusing or deferring physical therapy, missing clinic appointments, using prescribed medication improperly, not talking with their partners, and sexually "shutting down," that is, disengaging entirely from solitary or partnered sexual affects and activities. Moreover, their not infrequent disclosures that they "wouldn't even be at the clinic" if it weren't for their husbands suggested that women with vulvar pain bring a mix of desire (including for normalcy), verbal reticence, and bodily refusal to their (hetero)sexually disrupted situations.

To the extent that vulvar pain is a physiological realization of "actual distaste" (Frueh 2003, 139) and disparagement toward the vulva, it is possible to theorize penetratively prohibitive pain as the instantiation of a female (hetero)sexuality unsatisfied with commonly available sexual situations. Exhorted by the media—as well as their clinicians—to move beyond penetration and explore what else their genital and sexual bodies might enjoy, my informants routinely encountered male partners uninterested in such novelty. Vulvar pain patients normalized these interactions by keeping their own clinical focus on a restored tolerance for easy penetration, but it is here where I locate an unstable and inchoate ambivalence: stated desires were frequently not followed by problem-solving behavior, and patients who were able to engage in "successful" penile-vaginal intercourse sometimes described subsequent feelings of anger and resentment toward their partners ("Okay, you got what you wanted!"). Faced with disrupting the penetrative narratives through which their bodies are typically interpellated, many of these women maintained active investments in reproducing *and* resisting these narratives, rendering the option of sexually shutting down a sensible and perhaps more manageable one.

In exam rooms and in interviews, women described expectations and disappointments around sexualities that were constructed and overdetermined by mainstream discourses. Gynecological discourse and popular rhetoric compete and conjoin in writing so-called healthy sexual scripts that normalize a penetratively based heterosexuality, one that is serviced by a compliant vulva. But a vulva that doesn't "work," one, that cannot function as an enthusiastic (or at least tolerant) receptacle for heteronormativity, performs the cultural work of manifesting the female genital body in its entirety. I suggest that *this* sexuality remains inadequately theorized by feminist researchers—that its singularity is missed by theories dominated by both phallic and queer perspectives. A vulvar-based "sexual imaginary" (Gatens 1995, xiv) opens a space in which female genitalia can exist in all their corporeal potential—as labial, clitoral, perineal, and pelvic floor anatomy and sensation. Such an imaginary is not available to missing and alienated vulvas, locating women who recuperate their genitalia (e.g., through physical therapy) on the cutting edge of alternative female sexualities.

This sexuality is infused with possibility, with the carnal potential of a profuse, expansive, and largely untapped source of pleasure and female corporeality, with a "sex" that Irigaray (1985b) insists can never be just "one." One imaginary among many (Gatens 1996; Grosz 1995; Potts 2002; Segal 1994), a vulvar-based sexuality is one that women with vulvar pain are in a distinct position to inform. Queered by their marginal relationship to penetrative coitus, but materially and discursively invested in heteronormality, the bodies of many of my informants were sexually paralyzed by the impossibility of these contradictions. But feminist and critical theory that makes space for their experiences can unseat the assumptions upon which this stagnation rests, transforming an ambivalent vestibular refusal into a recoded and generative orifice, a window into the routine violence of heteropatriarchy. If we read the pain and burning of *vulvodynia* or VVS-afflicted genitalia as a way in to the conflicted desires, anger, and disappointment of (some) heterosexual women in the contemporary United States, we have established a new opening in sexuality studies through which to analyze the apparent investment that straight women make in penetrative coitus.

THEORIZING GENITAL PAIN

> My body in need of treatment and the productive society surrounding me are cast from the same mold.
> —Barbara Duden, *The Woman beneath the Skin*

Other Causes

I'm not convinced that a woman in the contemporary United States can escape the mediated and pernicious "blob" (de Zengotita 2005) of discursive contamination that I call genital dis-ease; indeed, if there is a clean or unpolluted cultural space in which the labia and vulva can take up residence, I remain unhappily unaware of its existence. Indeed, as recently as June 2012, Michigan state legislator Lisa Brown was ousted from her state's legislative chambers when she "failed to maintain the decorum of the House of Representatives" by using the word "vagina" in an abortion debate (Roberts 2012); and in 2014, a Japanese artist nicknamed Rokudenashiko (loosely, "good-for-nothing girl") was arrested on obscenity charges for distributing data from which 3-D models of her vulva could be printed.[10] When I interviewed women—in booths at Denny's or in bustling coffee shops—I sometimes asked them to ponder the physical space in which their words were being spoken, not just into the tape recorder on the table between us but into the air itself, the "open expanse" that Irigaray defines as "that [which] unfolds indefinitely and gathers all things together" (1993b, 40). I did this because I wanted us to imagine that our conversations—our public utterances of words and ideas too unsettling for legislative chambers and 3-D printers—were perfusing the space around us, seeping into the collective (un)conscious, by way of waitresses, menus, customers, and ambient noise. If the vulva needed to remain invisible in order for it to be culturally palatable, I thought, then perhaps our deliberate and unapologetic voicing, of both its existence and precarious state, might settle like so much dust onto the objects and people in its discursive and material circuits. Or that, like pheromones, our conversations could be naïvely absorbed through fluid and porous corporeal boundaries, influencing the instinctive behavior of those who were inadvertently exposed to them.[11]

Talking about their genitals is a behavior that is uniquely, though not exclusively, constrained for women with vulvar pain. Many patients told me that they had not discussed their symptoms with anyone but their partners and doctors. Others had confided in their mothers or another trusted intimate, but all agreed that their symptoms remained largely undisclosed to friends, coworkers, and relatives. In this section, I propose three "other causes of genital discomfort" (Harlow and Stewart 2003, 83) that structure the silence through which symptomatic women live their disease: genital dis-ease, unwanted genital experience, and genital alienation. Though each has the potential to exacerbate the severity of a woman's disease process, particularly if it keeps her from seeking treatment, they are also social processes capable of generating their own deleterious effects. I understand these "other causes" to affect women with and without vulvar symptoms, and I argue that their cultural and bodily impacts both precede and transcend the individual experience of pain. As social conditions contingent upon a particular set of historical and cultural variables (including patriarchy and heteronormativity), however, these "other causes" are also preventable, amenable to a host of cultural and political interventions that can help women rewrite the genital rules they have been handed.

Genital Dis-ease

In their important article "Clitoral Conventions and Transgressions," sociologists Lisa Jean Moore and Adele Clarke (1995) trace the clitoris's visual presence in—and absence from—anatomy textbooks from the twentieth century. Sampling a dozen books published between 1900 and 1991, these researchers demonstrate that graphical representations of the clitoris adhered to prevailing cultural discourses regarding female sexuality, with the clitoris disappearing (i.e., not being drawn) during periods when vaginal orgasms and penetration were prioritized in the medical and popular literature. Moore and Clarke's tracking of clitoral representation invites us to examine the relationship between social and medical discourses, as well as the role of representation in constructing anatomical—and therefore clinical—reality (Prentice 2012); how, in other words, "aesthetic and scientific paradigms, not empirical or experiential facts, determine understandings and even illustrations of genital anatomy" (Frueh 2003, 139). "Anatomies matter to feminists," Moore and Clarke insist, because they "create shared images which become key elements in repertoires of bodily understanding" (1995, 255). Bodily erasures, we can conclude, do not occur in a vacuum: if my genitals are missing from my doctor's textbook, they are likely missing from a wide variety of cultural locations with which that book intersects.

My own tracking of female genitalia has taken me to numerous field sites, including women's restrooms, popular television and film, physical therapy sessions, clinical and academic conferences, sexual health websites, exam rooms, undergraduate classrooms, feminist workshops, the local Planned Parenthood board, fiction and other literature, surgical suites, sex shops and in-home sex toy parties, political protests, and—most recently—the Texas House of Representatives during the 2013 antiabortion hearings that brought then–state senator Wendy Davis to international attention. In between these various sites, my fieldwork also takes place in conversations with other people, be they diagnosed informants, friends, students, or professional colleagues, and it is these dialogues and exchanges from which I discern the habitual and commonsense ways in which vulvar dis-ease is lived. Most recently, this involved an exchange with a longtime colleague who, after talking with me about some of the ideas in this book, suggested that maybe it wasn't "that big of a deal" for women to say "vagina" instead of "vulva." When I asked him why he didn't just call his wrist his hand or his thumb, given their anatomical proximity, he acknowledged that perhaps I had a point.

Inhabiting and observing these cultural spaces reveal that female genitalia are the subject of numerous forms of attention and intervention: "va-jay-jays" populate female-centric blogs and other forms of media; women get "vajazzled" by having their pubic hair replaced with patterned Swarovski crystals;[12] lists of the "Top 9 Most Amazing Vaginas"[13] and "10 Movie Vaginas Scarier Than the One in *Teeth*"[14] serve as clickbait for a number of pop culture websites (and only occasionally confuse vaginas with vulvas); and a variety of products target and commodify women's genital shame and insecurity, allowing them to sanitize their otherwise problematic privates in increasingly inventive ways. In 2010, consumers witnessed the debut of My New Pink Button, a temporary dye whose ad campaign, with its promise to "restore the 'Pink' back to a woman's genitals,"[15] renders anomalous any woman—particularly women of color—for whom rose-colored labia are not the norm. Once attuned to the implications of these diminutions and disappearances, it becomes almost impossible to ignore their importance, an orientation to pop culture that can attenuate one's enjoyment of otherwise female-friendly forms of entertainment. Indeed, this sensibility now extends to much of my social media circle, who fill my Facebook wall with vulvar-centered stories, most recently apprising me of an episode of the prison drama *Orange Is the New Black* in which a group of cis female inmates learn the details of their genital anatomy from their trans female peer.

These disavowing discourses, through which female genitalia are simultaneously named, disparaged, and erased, can also be tracked across numerous historical registers, "from Galen . . . through contemporary feminism" (Frueh 2003, 139). Aristotle, for example, equated bodily asymmetry with social hierarchy: women's "inverted" genitalia symbolized their overall inferiority (Laqueur 1990). Centuries later, biologist Georges Cuvier developed an intense fascination with the allegedly "over-develop[ed] . . . [and] disgusting[ly] deform[ed] . . . vaginal lips" (Fausto-Sterling 1995, 37)[16] of "Hottentot Venus" Saartjie Baartman; two hundred years later, another scientist named George (Henry), argued that the queer predilections of female "sex-variants" in WWII-era New York City were responsible for their "larger than average vulva[s]" (Terry 1995), relying on outdated and Lamarckian notions of bodily change to render the sexualities of these women deviant and culturally suspect. This thin slice of history demonstrates that vulvar dis-ease is tenacious but also highly flexible, evolving to accommodate any number of phallocentric and patriarchal social arrangements. Moreover, and as the work of Kimberlé Crenshaw (1991) and others has helped to make clear, vulvar disavowals always intersect with other cultural dynamics, including, respectively in these three cases, historical and geographical specificity, race, and sexual orientation.

The genital dis-ease deployed in these cultural spaces is varied, contradictory, layered, and recursive; the dirty jokes, sarcasm, and other forms of symbolic violence through which female genitalia are routinely disparaged index not only explicit misogyny but also ignorance and confusion regarding female anatomy and physiology. One final item from contemporary popular discourse, a fictitious profile of "hoo-ha specialist" Dr. Victoria Lazoff from the parodic newspaper *The Onion* (2009), expertly captures the shame and uncertainty that contour symptomatic women's (in)ability to describe their symptoms—the affective miasma they must negotiate to pursue and secure effective treatment. The piece also cleverly illustrates the ways that gynecological medicine can perpetuate, rather than clarify, our collective bewilderment:

> The world's foremost authority on ailments down south, Lazoff led a team of cutting-edge hoo-ha doctors to develop new strategies for detecting abnormal growth in . . . you know, that area. The accomplished physician humbly accepted medicine's highest honor . . . and spoke about the importance of regular screenings

to prevent unnecessary complications up inside one's business. "Recent advancements have brought us closer than ever to eliminating this threat, but early detection is still our best defense," said Dr. Lazoff, who earned a doctorate in lady parts from Johns Hopkins University.

In real-life gynecology, things are not uniformly better. In fact, they are sometimes worse, perhaps nowhere more evident than in the promotional materials of surgeons who specialize in cosmetic labial excisions. Labiaplasty, as the procedure is most frequently called, can be analyzed alongside several other forms of female genital augmentation, including "vaginal rejuvenation surgery" and the "G-shot," all of which purport to improve women's sexual lives. Though a thorough consideration of these procedures is beyond the scope of this book, it is important to clarify that vaginas and vulvas are targeted both separately and as a package, depending on the surgeon's marketing strategy. Labiaplasty, in brief, surgically removes portions of the labia majora (outside lips) that a woman, in consultation with her doctor, deems unsightly or otherwise excessive. Given my argument's focus, I am especially concerned about these reductions and about surgeons' routine use of the term "redundant" to describe the labial tissue marked for excision.[17]

Indeed, to fully appreciate the genital discomfort with which this book is most broadly concerned, we must consider the physical and social realities that are entangled with the emergence of these procedures. Indexing a convergence of feminist-informed bodily awareness, developing surgico-technical expertise, mainstreamed pornography, and what psychologist and activist Leonore Tiefer (2010) has called "retail medicine," labiaplasty offers shame-filled and sexually savvy female consumers the opportunity to have the vulvas of their—or their partners'—dreams. In one of the earliest and most eloquent analyses of this excisional phenomenon, cultural theorist Simone Weil Davis (2002) pointedly asks, "What do the aesthetics of a streamlined vulva signify? . . . In a world where many women have never thought about judging the looks of their genitals, even if they care about their appearance more generally, we should ask what criteria make for a good-looking vagina [*sic*], and who is assigned as arbiter" (13–14).

Unlike the fumbling and hesitant nature of genital pain diagnostic discourse, the medical condition of redundant labia is constructed and stabilized by the concomitant condition of genital dis-ease. Regular exposure to not just mainstream porn but also billboard and print campaigns exhorting them to "Feel like a woman again!" conditions women like Lori, whom I interviewed about her labiaplasty in 2010, to "start looking at [themselves and think] like 'Wow! (haha) Maybe, you know, I need it too.'" Her self-perception that her labia had "started getting bigger or saggier" led to a prompt evaluation by a cosmetic surgeon who assured her that, though in his estimation her vulva "wasn't that bad," he saw no reason not to comply with her request for a reduction.

Labiaplasty's increasing prevalence[18] (Braun 2005, 2010; Clark-Flory 2014; Green 2005) demonstrates that far greater numbers of women than those diagnosed with chronic vulvar pain sit in acute genital discomfort. And though many find their way to the waiting rooms of other clinics, hoping to tend to their bodily unease by having it neatly and professionally sliced away, others, including many of the women in this book, experience tremendous confusion over the role that this discomfort plays in their experiences of anorgasmia, disappointing or unwanted sexual encounters, or inadequately treated vulvar pain. Though the professional body of practicing gynecologists has spoken out against cosmetic labiaplasty (ACOG 2007), these intraprofessional differences are not clear to prospective patients. Moreover, confusion over vulvar etiquette is an unsurprising effect of being both a woman, who is encouraged to feel negatively toward her genitalia, and a patient, who is expected to care for them. Epidemiological data demonstrate that up to 40 percent of women with vulvar pain do not seek treatment (Harlow and Stewart 2005, 871), and I argue that this is—at least in part—due to the dis-eased tension between these two roles. Though it is true that most US women have been effectively conditioned to seek routine gynecological care (Bush 2000; Lewine 2014; Mehrotra, Zaslavsky, and Ayanian 2007), the scripts that structure the delivery of this care are written and managed by clinicians and not patients.

Heteronormatively oriented, gynecological encounters frame "real sex" in the same penetrative terms as do patients; in this context, vulvar pain is reported by patients as "pain with sex," a catch-all term that poorly approximates its anatomical specificity. Adhering to this script, including to her subordinate roles as both female and patient, significantly compromises the access that a symptomatic woman can gain to effective or expert treatment. The three words she needs to say—"My vulva hurts"—have likely never before passed her lips, nor those of her friends and acquaintances.

Physicians participate, therefore, in excisional discourse and genital dis-ease not only by performing surgical reductions but by limiting their clinical dialogue to the heteronormative concerns (contraception, pain with sex) that preclude a space through which a woman can discuss her external and nonreproductive genitalia—the clean space that this book posits and that might facilitate a simple description of "pain with sitting down." Though a gynecological encounter is technically a space where women are granted permission to speak plainly, many women find it awkward to move across the social registers that sanction frank references to their vulva and tell their doctor what they never tell anyone else. In their mainstream guises, medicine and "real sex" condition women to focus only on the parts of their genitalia that complement their respective penetrative goals: gynecological procedures and heteronormative intercourse (Kapsalis 1997; Laqueur 1990). Vulvar pain compels symptomatic women to inhabit more of their sexual bodies, but their genital dis-ease severely constrains their ability to broaden this confrontation beyond the immediate context.

Unwanted Genital Experience

A second cause of genital discomfort involves the high prevalence of sexual abuse and assault in the United States and the pervasive threats of harm to which women and their genital bodies routinely bear witness. Understanding these threats as compounded ones—that is, *of* violence and *to* bodily integrity—allows us to more adequately grasp the ways that vulvar pain and female sexual subordination amplify one another in the bodies of symptomatic women. Though managed by different types of professionals, sexual trauma and chronic genital pain index the often undisclosed nature of the female genital body. Laura Brown (1995), a self-described feminist psychotherapist, describes it this way:

> For girls and women, most traumas . . . occur in secret. They happen in bed, where our fathers and stepfathers and uncles and older brothers molest us in the dead of night [and] behind the closed doors of marital relationships. . . . These . . . are the experiences of most of the women who come into my office every day. They are the experiences that could happen in the life of any girl or woman in North America today. They are *experiences to which women accommodate; potentials for which women make room in their lives* and their psyches. (101; my emphasis)

Many of the women I interviewed reported pasts that included sexual abuse or molestation. Several had pursued counseling related to their experience and others were engaged in more self-directed recovery efforts. Some related their genital pain to these pasts and others did not. In the not-too-distant past, it was easy for clinicians to suspect—if not comfortably determine—that vulvar pain's inscrutability was related to a (buried) history of sexual abuse and to refer patients for counseling rather than investigate the physiological source of their symptoms. As I explain in chapter 4, sexual abuse and assault figure into the discomfort felt and perceived by women with *and without* vulvar pain. And as Brown makes eloquently clear, an "accommodation" to the possibility of sexual assault is "not outside the range" (100) of most women's experience. What I want to underscore is that, first, these private accommodations are also bodily—necessary corporeal adjustments to perceived threats of violence, threats that many feminist activists now consolidate within the term "rape culture," and, second, these routine and embodied tweakings are also made in reaction to the dis-eased discourses considered in the previous section.

A core argument of this book is that seeking relief for genital pain is hard work, and that the work done by these patients extends beyond the personal and well into the cultural. Indeed, I am suggesting that the experiential mode through which afflicted women first confront their symptoms is unavoidably tainted with cultural dis-ease, including the state(s) of genital risk engendered by what any man might do to her genitalia at any time. In the US legislative "war on women" that began during Barack Obama's first administrative term, the modes through which that threat could be imposed expanded significantly, coming to include the notorious transvaginal ultrasound probes to which women seeking abortions in some states are currently required to submit. In the years since, it has also come to include forms of online misogyny and violence to which female—particularly feminist—journalists and writers find themselves routinely subjected. Amanda Hess (2014) recounts the now-chronic states of watchfulness experienced by women like feminist columnist Jessica Valenti, who, like her peers, has been threatened with rape, murder, and a volume and scope of bodily harm previously unimaginable, though now grimly ordinary, on such a mass scale. Hess reports on the crippling lifestyle changes that the FBI advised Valenti to make: "to *leave her home* until the threats blew over, to *never walk outside of her apartment alone*, and to *keep aware of any cars or men* who might show up repeatedly outside her door. 'It was totally impossible advice,' [Valenti] says. 'You have to be paranoid about everything. You can't just not be in a public place'" (Hess 2014; my emphases).

With a nod to Raymond Williams (1978), Sara Ahmed (2014) calls Valenti's paranoia a "feeling of structure" and argues that "feelings are how structures get under our skin," curbing our abilities to be full economic and political citizens. Echoing Brown (1995), as well as prominent feminist blogger Melissa McEwan (2014), Ahmed insists that though sexualized misogyny—online or otherwise—takes an enormous affective toll on its targets, feminism is fueled by its "willing[ness] to venture into secret places of pain." And though my politics are sympathetic to these arguments, I want to add that threats to genital integrity are not *always* in the form of penetrative or sexualized violence and that they are not always perpetrated by men. Cultural disparagement comes in many forms, and in my various guises as friend, health care provider, colleague, anthropologist, professor, and feminist observer, I can report—and only partly explain—the variety of ways that women both produce and sustain the states of pollution and excess through which their bodies are socially grasped. And though long accustomed to scribbling it down in a field notebook as soon as I can, I still wince with confusion (and some despair) when it is women who participate; when they tell me, for example, about a friend of theirs who always "said that a vagina [sic] looked like something that got dropped out of a ten-story window" and then wait for me to laugh.

I use the phrase "unwanted genital experience" to signify this more collective mode of female sexual subjectivity, the mode through which any of us might react to this joke, the political salience of the transvaginal ultrasound, or a set of before and after labiaplasty images. "Many women," argues Brown, "have never been raped [yet] have symptoms of rape trauma . . . [—being] hypervigilant to certain cues, avoid[ing] situations that [they] sense are high risk, go[ing] numb in response to overtures from men that might be friendly" (1995, 107). Here, it is the *bodily* aspects of Brown's assertion to which I want to pay heed, because I am arguing that just such behaviors occur in response to the vagina-out-the-window joke, at least the first time it gets told. And though discourses of devaluation are less physically threatening than acts of sexual violence, I suggest that the differences are of degree rather than kind. It is easy for most of us to understand the bodily hypervigilance of a sexual assault survivor; we might also readily appreciate that a woman who has experienced repeated pain on genital contact might develop the same set of protective maneuvers. But I am asking how far we need to stretch our (feminist) analysis in order to imagine that a woman whose genitalia have been routinely and unquestioningly insulted might *also* come to incorporate this behavior.

Sociologist Pierre Bourdieu (1977) describes a bodily hexis, "imperceptible cues" (82) and a "pattern of postures" (87), that both structure and are structured by the conditions of one's social world. The hexis, says Bourdieu, "speaks directly to the motor function" of a body that is "charged with a host of social meanings and values" (87). In the contemporary United States, women experience far more shame than pride, and more fear than joy, as they live *with and in* their vulnerable bodies. Vulvar pain and sexual assault can exacerbate this physical reality, but they do not create it; consistent and ideologically driven devaluation of the female sexual body, however, does. Contractions of the pelvic floor, the reactive "tail-tucking" under consideration here, can be clinically measurable in the bodies of women without histories of pain or assault but whose lives contain "other causes" of genital discomfort, such as the cultural and political subordination of one's bodily autonomy. These bodily changes, Bourdieu argues, are "political mythology realized, *em-bodied*, turned into a permanent disposition, a durable manner of standing, speaking, and thereby of *feeling* and *thinking*" (1977, 95; emphasis in original).

In bringing sexual assault, genital pain, and disparaging discourses under one analytic umbrella, I am suggesting that the female corporeal situation is rife with unwanted genital experience. US women respond to demeaning folklore and sexual violation through an enormous range of sensibilities and practices, many that directly contradict the shame and self-censorship that I propose here (Johnson 2002; Potts 2002; Segal 1994). But this book offers compelling evidence that some female bodies absorb these unwanted experiences and transform them into physiological and discursive states of alienation. Indeed, the three bodily modes that Brown singles out for our attention—hypervigilance, avoidance, and a state of numbness—are prescient descriptions how vulvar pain is often lived, with numbness being the desired outcome of one treatment modality. I join Brown in suggesting that many more women than those marked by disease or a history of assault live their genitals through any or all of these states. In making room for jokes, innuendo, and insulting discourse, cultural situations to which female bodies might protectively and shamefully respond, I seek to widen the parameters of what constitutes unwanted genital experience.

Genital Alienation

Profound levels of alienation are the risk, if not the routine reality, of the conditions examined in the previous two sections. I therefore use the term "alienation" to describe a spectrum of distaste and ignorance, moored by the absences of silence and erasure at one end and by the (hyper)presence of pain, pornographic amplification, and felt excess on the other. Without a clean space in which they can have a vulva, women oscillate between and within two unacceptable alternatives: a no-space and a contaminated one. In an invisible no-space, the vulva simply goes missing—absent from conscious awareness, untouched for its own sake, and attended to only by others (e.g., providers)—and I borrow the concept of agnosia from neuropsychology to describe this embodied modality. In the context of a lingering phantom pain that is felt where an amputated limb once existed, agnosia can be understood as its opposite, that is, "the nonrecognition of a part of the body as one's own" (Grosz 1994, 89; see also Sacks 1987). This state is not constituted by a simple reluctance to touch or examine one's genitalia; rather, this is the space through which a woman remains clitorally anorgasmic, verbally unable to describe her genital anatomy, and perceptually unable to recognize a visible or palpable labial lesion. A contaminated space, on the other hand, is manifest through the consumption of labiaplasty and Brazilian bikini waxes, where women confront their genitalia by removing the felt excesses of labial tissue and too much pubic hair. But the excessive vulvas with which this book is most analytically invested are those that are in pain. Reddened, itchy, "on fire," and recalcitrant, these labia are amplified and simultaneously muffled by bodies that have been discursively disciplined to keep them quiet. In order to get better, they demand a quality of attention incompatible with either of these alienated—and embodied—extremes.

The work of Elaine Scarry (1985) and others (Jackson 1994; Leder 1990) has demonstrated that pain has alienating qualities all its own, often able to transcend, in this case, the awkward intimacy with which many women encounter their symptoms. But with vulvar pain, we can observe what Blackman (2010, 5) calls the "singular-plural nature of personhood": how various states of alienation inform each other in the patient's *individual* body, as well as how levels of *social* alienation articulate with those that are personal. Genital pain's volume is typically high enough to be perceived by the woman, her partner, her provider, and any other ears invested in her bodily well-being. But her pain is simultaneously absent from many clinical—and most social—registers, resulting in a poor fit between her symptoms and society's response to them. Dedicated vulvar specialty clinics constitute the kind of therapeutic hyperpresence demanded by these conditions, but, as this book will make clear, treatment strategies are typically less than adequate and most patients continue to have significant amounts of pain while under an expert's care This bodily reality leads to yet another level of alienation, as women whose pain exceeds their personal thresholds frequently choose to avoid the situations that threaten its recurrence. For most of the patients that I met during my fieldwork, this meant shutting down their sexual bodies and selves—a strategic and self-directed "numbing" intended to prevent an alienating amplification.

These fluid movements across and between various states of alienation are well characterized by Elizabeth Wilson's (2004) reworking of the Freudian concept of obligation. Located in Freud's writings on neurasthenic melancholia, and deployed in her own work to analyze mind-body relations, Wilson's concept of dynamic obligation displaces and obviates linear (or cause-and-effect) explanatory models without dispensing with a binary that is often in tension: "Freud's use of obligation . . . denatures the human- and conscious-centric sense with which obligation is used elsewhere . . . [and offers instead] one way of understanding a relation between psyche and soma in which there is a mutuality of influence, a mutuality that is interminable and constitutive" (22). I return to this theme of obligation repeatedly throughout the book in order to account for the seeming ease with which vulvas are disavowed—made present so that they can be made to disappear. In the contemporary United States, women care for their sexual bodies in curious ways: removing their protective pubic hair to better display their labia, surgically trimming erotic tissue in search of greater genital pleasure, and creating greater distance between their sexual anatomies and their menstrual and contraceptive habits (Houppert 2007; Muscio 1998). In this genital hexis, absence, presence, agnosia, and amplification "map [each other's] strange elasticity" (Povinelli 2006, 9) as they find themselves socially and somatically *obliged*.

OVERVIEW

In her critical examination of the journals of an eighteenth-century German physician, historian Barbara Duden (1998) writes,

> The first step toward understanding the complaints of the women of Eisenach was . . . to realize that my own certainties about the body are a cultural bias, one which perhaps I could even learn to transcend. I had to create some distance to my own body, for it was clear that it cannot serve as a bridge to the past. (vii)

I take Duden's conclusions to heart and introduce this book within the context of similar distancing efforts. I have thought through the lived experience of vulvar pain from numerous angles, some of which are worked through in the coming chapters and some that have since been discarded. Underlying most of them, however, has been an attempt to keep a social constructivist position in dynamic dialogue with the beliefs and attitudes of patients, physicians, partners, and the researchers with whom I opened this chapter, the actors invested in the physiological reality of vulvar pain. I have hesitated to attach too much biological reality to symptoms that I believe to be profoundly social, and I have wanted to interpret these bodies symbolically at the same time that I have tried to explain physiologically intractable pain by deepening the perspectives of embodiment theorists. And I have done all of this without knowing—in my own body—what it feels like to genitally reject the penetration or approach of my partner, my doctor, or my own hand. As an anthropologist and critical theorist, I have used the tools at hand to delineate the arbitrary social conditions through which vulvar pain is experienced. In doing so, I denaturalize not only the bodies of my informants but also my own, other women without pain, and—ideally—any body that can be interpreted through the theoretical perspectives offered here.

Via a corporeal metaphor, in which the vulva moves from an insinuated event to a generative force, this book establishes the material fact of female genitalia. Each chapter is an attempt to instantiate one or more of the embodied states through which these pain conditions are both encountered and lived. I suggest that the symptoms arise in the body as they arise in culture; in Duden's words, these pain conditions are an "incarnation of the world in the body" (1998, 38). My metaphor, then, is an attempt to map the female body that I have begun to theorize. I argue that this body has accumulated far too much unwanted genital experience and that this burden makes itself manifest through a sense of shame and felt excess and a disavowed and contaminated absence. This body has learned to strategically erase itself from so-called civilized discourse, compromising its ability to integrate into its social world, yet retains the potential to generate novel and unpredictable sexual morphologies.

Chapter 2, "Examination," provides much of the factual information needed to read the book. I describe the details of my research project—including patients, research site, key informants, and the research methods I employed; I also describe the medical conditions themselves in greater detail. Further, I use critical race theories to analyze the demographics of women diagnosed with vulvar pain, the majority of whom are white, educated, heterosexual, economically stable, and insured. In chapter 3, "Accumulation," I detail the variety of items collected by women with vulvar pain: disparaging discourses, years of pain, sexual disappointment, marital discord, and drawers-full of ineffective medication. I theorize accumulation as dynamic layering—where personal resources and socially structured realities shift, settle, and erode in individual bodies that move in and out of relief-seeking behavior. I also introduce the concept of genital "baggage" and question the ease with which it can be discarded or collected.

In chapter 4, "Manifestation," I unpack the ways that vulvar disease "shows up" in a variety of clinical spaces, including a major medical conference about vulvar pain. Here, vulvar pain is theorized as an eruption—a raw and uncomfortable presence on the skin's surface as well as on the landscape of gynecological medicine; I argue that what manifests in both cases is obliged, in the Freudian-Wilsonian sense, to the accumulated discourses described in the previous chapter. Much of chapter 5, "Integration," takes place in physical therapy sessions, and I argue that the intimate nature of this treatment compels diagnosed women to confront their genitalia in novel ways. I introduce the concept of "missing" vulvas and discuss the utility of the neuropsychological concept of agnosia in understanding these unrecognized genitalia. Agnosic nonrecognition is inconsonant with the goals of physical therapy, and women who can fully participate in this treatment option get better in ways that exceed the results of surgery or pharmacotherapy alone. Ethnographic data detail this process and also delineate the various personal and social barriers that prevent some women from accessing physical therapy.

Chapter 6, "Generation," elaborates a model of bodily and sexual generativity; via the work of neurologist Oliver Sacks, I theorize disease processes as sources of bodily limitation *and* potential. The notion of generativity is furthered with an analysis of the responses to an interview question: "What do you think that this disease/experience has given you?" Ranging from "Absolutely nothing," to "Only grief," to extended meditations on patients' maturation as individuals, partners, role models, and survivors, patients' reflections provide the bulk of the chapter. The women's own words—in lengthy excerpts, as organizing themes, and in dialogue with the work of corporeal feminists—are used to describe a female sexual subjectivity with an affective ambivalence and an inchoate autonomy. Finally, in chapter 7, "Evaluation," I discuss the cultural work performed by women with vulvar pain. I argue that afflicted women evince a generalized pattern of (hetero)sexual distress and demonstrate the modes through which noncompliant vulvas unsettle patriarchal gender norms. I conclude the chapter by asserting that the culturally charged vulva of my argument provides new concepts through which to think through genital/sexual health and well-being.

With these chapters, I seek to "close the distance between the body and the world" (Duden 1998, 38–39) and immerse the reader in the experience of a profoundly paradoxical pain. Vulvar pain is acutely felt and amplified at the personal level—it is dermatologically and neurologically loud; it "shuts down" a symptomatic woman's sexual possibilities; it is red and itchy and narrated to physicians (and husbands) as "don't go there." But, like the clitoral orgasms of a woman with *lichen sclerosus* (see chapter 2), it is simultaneously muffled, erased, and all-but-ignored at the broader—and collective—levels of cultural discourse and institutions. The chapters before you are meant to convey these seemingly contradictory states of existence and offer a glimpse into the paralytic torpor these states sometimes produce. But they also posit another state of existence, a potential way out of the condition that I refer to as Vulvar Disease. This is the pathway chosen by patients that "got better" during my time in the field—usually through a form of physical therapy that facilitated an acceptance and negotiation of the extracorporeal aspects of their pain, whether these be emotional or cultural, based in the past or drawn from the present. Vulvar pain conditions are infused with moments, histories, and discourses through which the female genital body is made to disappear. By making this body appear—*again and again*—on these pages, I insist on its legitimate cultural recognition and its rightful place in the body images of the vulnerable and dis-eased women who are at the heart of this work.

2

EXAMINATION

Clinical Interpretations of Vulvar Pain

During thirteen months of fieldwork, I developed relationships with two physicians, a handful of local physical therapists, one nurse, several medical assistants, a dozen gynecology residents, a half-dozen medical students, a sex therapist, and forty-two women who allowed me to gather the details of their disrupted lives. I met and observed the clinical consultations of many more (close to one hundred), but these forty-two made time to meet with me outside the hospital and to talk at length about their struggles with symptoms, sexuality, and genital well-being. I observed their surgeries, bought them dinner, accompanied them to the pharmacy and physical therapy, brought them to yoga, had meals in their homes, brought them cake and flowers, got drunk (with one), and listened attentively to stories that, in their words, had never before been told in their entirety. To anyone. And though I socialized less with the care providers, my ubiquitous presence in their exam rooms, surgical suites, and treatment sessions led to hours of conversation, much of it refreshingly analytical. This mix of methodologies allowed me to directly observe a developing and deepening set of investments in the physiological reality of unexplained vulvar pain. My methods also enabled patients to tell stories that complicated these investments, often significantly, leaving me confused at times about how to best tell them to a wider audience.

Like all modes of inquiry, fieldwork articulates with histories and contexts that transcend our ethnographic encounters and shape our analytic insights. This chapter locates vulvar pain conditions within these wider worlds in order to foreground how these diseases have been and continue to be imagined by clinicians and researchers. I begin by explaining the disease conditions most commonly managed during my fieldwork, followed by a focused history of how, as diagnostic categories, they have produced and reflected particular assumptions about race, gender, and the female (sexual) body. I conclude with brief descriptions of the clinic itself, including the research hospital that housed it, and the medical and administrative personnel that kept it running.

CATEGORIZING THE CONDITIONS

> *Vulvodynia* is chronic vulvar pain in the absence of objective abnormalities such as infection or dermatoses. Dysesthetic *vulvodynia* (newly termed generalized vulvar dysesthesia) refers to episodic unprovoked stinging, burning, irritation, pain, or rawness anywhere on the vulva. *Vulvar vestibulitis* (newly termed localized vulvar dysesthesia) refers to pain consistently localized by point pressure mapping within the vulvar vestibule. (Harlow and Stewart 2003, 82; my emphasis)

At first glance, this passage appears to provide a straightforward distinction between the two most common forms of vulvar pain. But in their revisions of extant nomenclature (see figure 1), these definitions also index the inherently dynamic nature of disease conditions and the ways that genital pain is continually realized through medical practice and symptomatic bodies. Definitions and diagnostic criteria for vulvar pain are many things: singular (pain at the vulva) and multiple (provoked vs. unprovoked? burning vs. knifelike? redness, rawness, irritation, or none of the above?); clinical (neuro-inflammatory

FIGURE 1. Vulvar pain nomenclature

Vulvar Pain	
Local and contact-specific	Generalized and diffuse
Vulvar vestibulitis syndrome	Vulvodynia
Localized vulvar dyesthesia	Dyesthetic vulvodynia
Localized provoked vestibulodynia	Generalized unprovoked vestibulodynia
Vestibulodynia	

pain) and sociological ("something of a medical mystery"; Kaler 2006, 82). They are also a reflection of current trends in medical practice, scientific inquiry, and funding streams. The authors of the consensus panel convened by the National Institutes of Health (NIH) in 2004 defined vulvodynia as "chronic pain lasting from 3 to 6 months in the vulvar region without a definable cause" (Bachmann et al. 2006, 448) and *vulvar vestibulitis syndrome (VVS)* as a diagnostic "subset" of that condition. More recently, both conditions have been collapsed into one term—*vestibulodynia*—which can be subcategorized into provoked versus unprovoked and generalized versus localized forms (Bornstein et al. 1997; Edwards 2004; NIH 2012; Tuma and Bornstein 2006). In this section, I use these clinical descriptors as background to my ethnographically informed depictions of vulvar pain and suggest that neither is more accurate or useful. Rather, I use the stories of women I met in the field to animate the medical narratives that structure the institutional experience of vulvar pain.

During my fieldwork, patients at the Vulvar Health Center (VHC) were diagnosed within one of three disease classifications: (1) *vulvar vestibulitis syndrome (VVS)*, (2) *vulvodynia*, or (3) one of three *lichens—lichen planus (LP)*, *lichen simplex chronicus (LSC)*, and *lichen sclerosus (LS)*. Most clinical literature excludes the latter set of conditions from the definition of chronic vulvar pain, because their etiologies and treatment trajectories are more clearly defined. As a vulvar *health* center, however, the VHC managed patients with these syndromes, since the experience of living with them overlaps significantly with those that are more "mysterious" (Labuski 2011). Indeed, the phone calls that led to an initial appointment at the VHC were suffused with a genital alienation that did not discriminate between poorly and well-understood conditions, and all three lichens were seen and managed routinely by the two physicians with whom I worked, Drs. Robichaud and Erlich.[1]

Vulvar Vestibulitis Syndrome

VVS is distinguished from other categories by its precise and singular location. The vulvar vestibule is a small (0.5–1.5 cm), horseshoe-shaped area of endodermic tissue surrounding the opening to the vagina (the introitus). Viewing an introitus as the center of a clock face, the vestibule is the adjacent and typically paler skin that extends outward from the 3:00 to the 9:00 positions (see figure 2). Though variations are not uncommon, women with a "classic" presentation of VVS experience pain only in this area (localized) and only with touch (provoked).[2] The pain has long been thought to be neuropathic, inflammatory, and superficial in nature (Bachmann et al. 2006; Bergeron, Binik et al. 1997; De Andres et al. 2015);[3] for this reason, the application of a topical anesthetic (e.g., lidocaine—in liquid, ointment, or gel form) will usually eliminate the pain for short periods of time. Clinicians diagnose VVS by first localizing the woman's pain to the vestibule and then applying light pressure with a cotton swab from 3:00 to 9:00, gauging both the intensity (0–3, or 1–10) and quality (burning, zingy, raw) of sensations (see figure 3). Once these parameters have been established, the clinician covers the entire area with liquid lidocaine, waits several minutes, and then repeats the "Q-tip test." A reversal or extreme reduction of pain is diagnostic for VVS (Ventolini 2011).

FIGURE 2. A clinical vulva

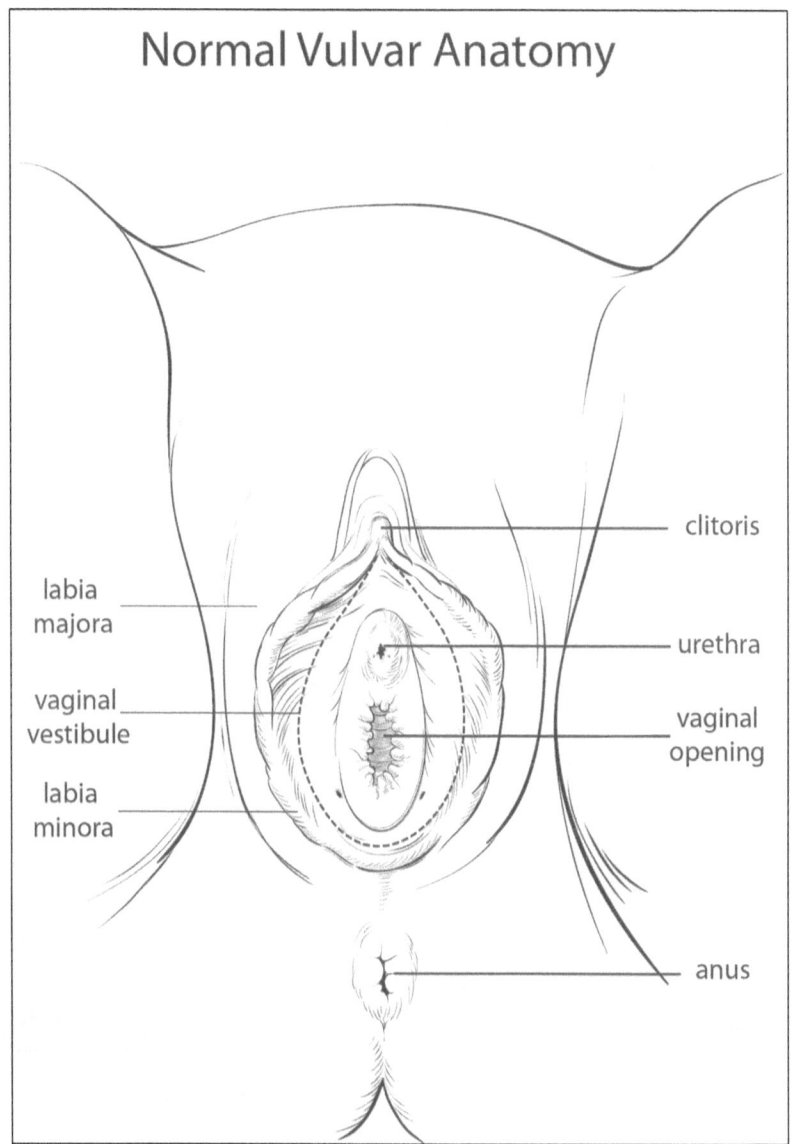

Image used with permission of Robin Jensen.

FIGURE 3. The cotton swab, or "Q-tip," diagnostic test

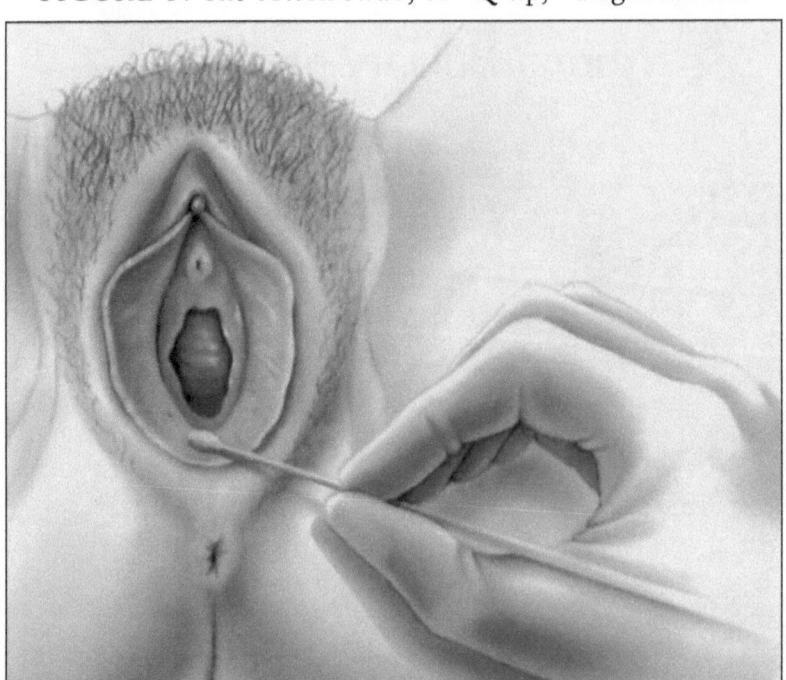

Source: Haefner, Hope K. 2000. "Critique of New Gynecologic Surgical Procedures: Surgery for Vulvar Vestibulitis," *Clinical Obstetrics and Gynecology* 43(3): 3. Used with permission.

At the VHC, Drs. Robichaud and Erlich typically asked patients to rank their pain from zero to three (with three signifying the worst), though they occasionally switched to a zero to ten scale; they supplemented these numeric gauges with words and phrases—such as "zingy" and "just kinda 'ouch'"—that helped to fine-tune their clinical impressions. Both physicians also invited patients to provide their own descriptors, soliciting what Kempner calls the "embodied epistemolog[ies]" (2014, 91) of people afflicted with a disease condition. Frequently, either clinician would provide a patient with a handheld mirror and ask her to visually identify the "zones" where her pain began and ended. In the majority of exams that I observed, the woman had neither a working knowledge of nor a language with which to describe her vulvar anatomy; these same women, however, were able to be very clear about what hurt and what didn't. "Yeah, right there! That does not feel good!" was a phrase I heard over and over again while clinicians mapped out a woman's vestibule. These women were also usually able to reach down to their genitals with their fingers and orient the clinician at the beginning of the exam—"Um, it's usually right around here . . ."

In order to be clinically diagnosed, then, patients needed to enact or "articulate the complaint specific to" (Mol 2002, 43) VVS; in this case, a woman was expected to display a change in her response to the light touch of a Q-tip as it moved from an asymptomatic area (such as her inner thigh) to her vulvar vestibule (Bachmann et al. 2006). Depending on the amount of pain she displayed, a clinician might also ask her to supplement the exam findings with an appropriate symptom history, one that "open[ed] up" (Mol 2002, 42), rather than foreclosed, the diagnostic trajectory. Though often significantly alienated from their vulvas, and therefore unable to apprehend the localized nature of their pain, the patients I saw almost always described their symptoms in relationship to genital contact, particularly penetration delimited to tampons, speculums, and penises.

The VVS patients from my sample were all heterosexual (married or engaged), insured, and college educated (or in the process); all but one were Anglo-American (Mira was South Asian.) Some came to the clinic to see Dr. Erlich specifically, in order to request a surgical procedure with which she had considerable expertise; some came hoping to find out that their condition was less serious than they feared; and some worried that their husbands would leave them if they could not access treatment that made penetrative coitus possible. All but one were religious and had postponed sexual relations with their partners until they were married or, in Mira's case, engaged. All were in their "childbearing" years (between menarche and menopause), and most were under thirty at the time of their diagnosis. Some discovered their pain in their first effort to use a tampon, and others became aware of it with, or shortly after, their first attempts at vaginal intercourse. All of these narrative events and demographic factors are consistent with the clinical literature, right down to the details of "disastrous" wedding nights and the overconsumption of alcohol in hopes of relaxing their (perceived) sexual inhibitions (Buchan et al. 2007; Kaler 2006; Munday et al. 2007).

Treatment options for VVS were directed toward anesthetizing, repairing, or excising affected skin. All patients were offered a prescription for the liquid lidocaine used diagnostically by the physicians and were encouraged to use it liberally in their attempts to engage in penetrative activities. Each patient was oriented to her vestibule with both a mirror and an overscale drawing (figure 2) and was instructed to apply lidocaine to the indicated areas five to ten minutes before she anticipated contact (including physical therapy sessions and clinical exams). Dr. Robichaud also routinely prescribed the nightly application of a lidocaine ointment, based on a study that had demonstrated some reduction in overall pain after seven weeks' use (Zolnoun, Hartmann, and Steege 2003). Beyond anesthesia, patients were counseled through various configurations of other available therapies: pharmacological, physical, laser dye, and surgical.

Both vulvodynia and VVS are understood in neuropathic terms, but a primary difference in the medical narratives popular during my fieldwork was that VVS's pain was local, meaning at the peripheral nerve endings, whereas vulvodynia was generalized, or based in the central nervous system. This distinction has become clearer through the limited response that VVS patients have shown to the use of tricyclic and SSRI antidepressant medications, which target the regulation of neurotransmitters like serotonin, norepinephrine, and dopamine.[4] In other words, regulating the circulation of neurotransmitters can affect both mood stability and—perhaps—some kinds of systemic pain perception; the localized and inflammatory nature of VVS pain, however, is not receptive to drugs that behave in this way. For this reason, SSRIs were occasionally offered to patients thought to have a vulvodynia component to their pain or to women whose pain had become enmeshed with depressive symptoms, but it did not, at the time of my fieldwork, constitute a first-line therapy for VVS.

Many clinicians—including Drs. Robichaud and Erlich—believe that the skin of patients with VVS has been locally, superficially, and situationally damaged through a combination of genetic predisposition and exposure to environmental stressors (such as repeated yeast infections or viral exposure) (Chadha et al. 1998; De Andres et al. 2015; Foster et al. 2007). Leclair et al. (2007) describe the histologic features of VVS as consisting of "nonspecific inflammation of the vestibular epithelium and a higher density of nervous tissue" in addition to areas of angiogenesis (i.e., construction of new blood vessels) that are "thought to be a result of vascular injury, although the nature of the insult is unclear" (54). Their article further addresses the (at times) successful efforts to attribute an infectious, hormonal, and immune system etiology to *VVS*, although they conclude that because the "significance of these changes is undefined, . . . it is unclear what part [they] play . . . in the overall pain syndrome" (54). Not wishing to limit their treatment efforts by an incomplete causal narrative, however, the doctors at the VHC directed their interventions toward the removal or alteration of the adversely affected skin of VVS patients.

Physical therapy (PT) was also prescribed for VVS, though only as an adjuvant treatment. This is because women with VVS have learned to flinch, tense up, and pull away from anything resembling vulvar or vestibular contact; their pelvic floor muscles have often become tight and contracted, leading them to feel pain and "burning" sensations with penetration, regardless of whether their vestibular skin has been removed or repaired. PT attempts to resolve this concomitant problem but is not believed—at least by most clinicians—to directly address the "skin pain" of VVS (Bergeron et al. 2002; Goetsch 2007).

Dr. Robichaud always prescribed lidocaine, almost always prescribed PT, and occasionally treated patients with an SS(N)RI. At the time of my fieldwork, she did not consider surgery to be a treatment of last resort, and she discussed it with patients at their initial consultations. However, she also assertively encouraged her patients to take a "long view" approach to their symptoms, suggesting that they first try her other suggestions for three months. Dr. Erlich, on the other hand, accumulated much of her vulvar expertise around the development of a surgical technique, and patients often came to the VHC for this very reason. Her treatment plans were therefore less conservative, and she was reluctant to withhold surgery from a woman for whom it was her strong preference.

Both doctors also offered their VVS patients noninvasive laser therapy as a middle-of-the-road approach. Laser therapy was thought to address the angiogenetic element of VVS—targeting and selectively disrupting the increased blood vessel formation—as well as the nerve density component, while still "preserving [the] anatomy" (Leclair et al. 2007, 54) that surgery excised. Only a small percentage of the VVS patients I met opted for this procedure: four to seven treatment sessions were recommended before efficacy could be evaluated, and many women wanted the quicker and more "guaranteed" solution promised by surgery's increasing success rates (Bergeron, Bouchard et al. 1997; Goetsch 1996; Kehoe and Luesley 1999). Additionally, since the use of pulse/dye lasers was still experimental, some women worried that their off-label use would not be covered by their insurance companies.[5]

Finally, both doctors offered and performed a procedure known as a modified vestibulectomy.[6] Surgery was typically chosen fairly early in a patient's treatment plan, and it was rare that I saw someone move through all the other options before electing to undergo a vestibulectomy. The procedure consists of mapping the skin with a Q-tip test (see figure 3) and then excising the entire affected area without excessively disrupting genital anatomy. Guided by the pain reversals produced by topical anesthetics, the excision is only two to three millimeters deep; skin from residual hymenal tissue is then pulled down from the introitus to cover the open area (see figure 4). Surgery was sometimes described to patients in both curative and preventive terms: the ectodermic nature of hymenal tissue was thought to obviate the possibility of a recurrence, since only endodermic tissue, in Dr. Erlich's words, could "get vestibulitis." Excised tissue was always biopsied, although the results did not consistently demonstrate an inflammatory or pathological disease

FIGURE 4. A modified vestibulectomy

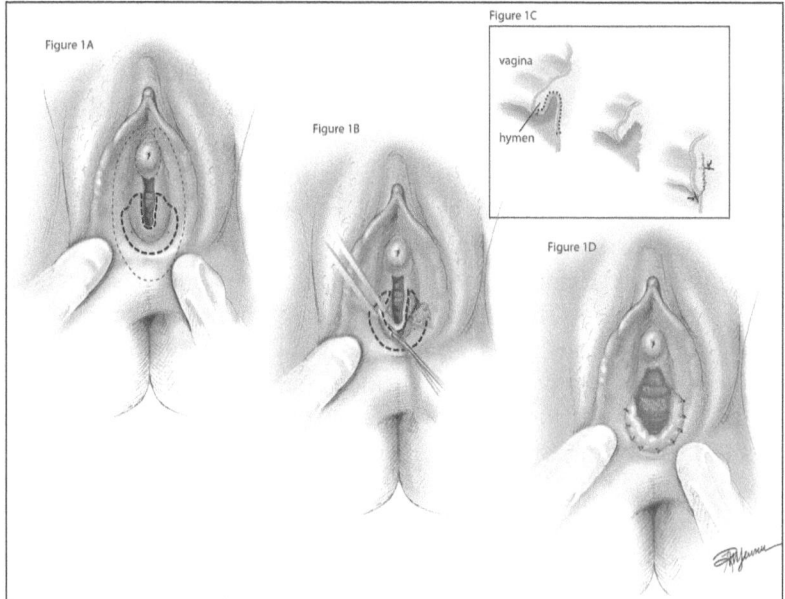

Image used with permission of Robin Jensen.

process. All surgical patients were highly encouraged to pursue PT and sexual/relationship counseling, as well as to practice vulvar care measures with their skin (i.e., hypoallergenic products, nonirritating fabrics). During my fieldwork, I followed five women through surgery and learned of only one postoperative complication (one that resembled those described in the literature), which was eventually resolved.

Still controversial, and a last resort for some clinicians, surgery necessarily involves some risk—infection, excessive bleeding, abnormal scarring or other anatomical complications, and continued or worsened pain—and these are amplified as well as downplayed in the literature (Brokenshire, Pagano, and Scurry 2014; Goetsch 1996, 2009; Tommola, Unkila-Kallio, and Paavonen 2011). As with other forms of genital cutting, including those performed for religious and cultural reasons, and on intersex and trans individuals, there is scant attention paid, at least thus far, to the potential long-term consequences of vestibulectomy, particularly regarding whether and to what degree it influences genital sensation and orgasmic capacity. Because vestibular excisions are superficial and tightly circumscribed, surrounding nerves and neuroanatomical tissue are believed to be unaffected, though this has neither been demonstrated nor explicitly addressed in the literature. Likewise, and as Katrina Karkazis (2008) has shown in her research with intersex infants and families, many surgeons doggedly "assume that innervation is tantamount to erotic sensation" (164), making clinical research regarding the long-term erotic capacities of these patients unlikely; indeed, given the sexual complications surrounding vulvar pain, assessing such outcomes may be especially vexing. But we should bear these complexities in mind. Gillian Einstein (2008), for example, argues that a better understanding of genital cutting's neurobiological correlates can help to "elucidate a biology of human difference" (85), a goal shared by a host of trans and feminist science scholars whose work questions the links between genital anatomy, gender, and imagined bodily capacities (Karkazis 2008; Spade 2006). In this broader context, vestibulectomies can and should be located alongside other forms of cutting as ripe for feminist and queer biological investigation.

Dysesthetic Vulvodynia

Vulvodynia, or generalized vulvar dysesthesia, was the second category of pain managed by the clinicians at the VHC. Despite their ongoing evolution, the various terms used by researchers to refer to dysesthetic vulvodynia (*DV*) loosely translate into *maladaptive or inappropriate pain sensations at the vulva*. Like their cluster of symptoms, the characteristics of women diagnosed with this condition are slightly less circumscribed than are those of VVS patients, almost all of whom narrate a predictable and similar set of events leading up to their consultation. In contrast, women with DV, though phenotypically and socioeconomically similar (to one another and to women with VVS), are slightly more varied in their disease presentation, sexual histories, responses to treatment, and personal narratives. At the VHC, these women were typically older than VVS patients—the youngest in my sample were at the end of their childbearing years—and several were less educated. This group also contained more outliers (e.g., one woman in my sample identified as a lesbian) but was generally in line with the prevalence data in that almost all of the women that I met with DV were Anglotypic and insured (Arnold et al. 2006; Baggish and Miklos 1995; Edwards 2003; Gordon et al. 2003).

The two other significant differences between this group of patients and women with VVS were the characteristics of their pain and their relationships to penetrative intercourse. All of the DV patients I encountered developed their symptoms *after* a period of time (up to ten years) when sex had not been painful; some were in sexual relationships that were now hampered by their symptoms and others were not. These characteristics are related to a third distinction: because the pain of DV is unrelated to touch or provocation, these women were far more likely to seek relief on their own (rather than a partner's) behalf. DV pain is more diffuse than VVS—it cannot be precisely mapped or delimited to one or two areas of the vulva, and the quality is more fluid. It can be constant, flare-like, dull, burning, cyclic or predictably aggravated, concomitant with another condition, responsive to a wide variety of products, or all (or none) of the above. A so-called classic presentation of vulvodynia includes pain that is prohibitive of *something*—this means that the removal or cessation of particular fabrics, products, and activities often constitutes a significant part of a DV patient's life. In the northwestern United States, this was especially notable, as roughly half of the patients I met rode horses or bicycles more than casually and had to either give up or drastically alter their relationship to these activities. Julia Kramer, a young Latina whose pain had both VVS and vulvodynia components, told Dr. Robichaud that her drawer of non-cotton lingerie had become "a museum"—full of panties that she continued to admire but could no longer wear. It is important to reiterate that none of these irritating agents or activities are thought to cause DV; indeed, its onset is most often unrelated to any identifiable event or causative agent (Buchan et al. 2007; Munday et al. 2007).

The uncontained nature of vulvodynia seeps into the diagnostic process itself in that DV is often a diagnosis of exclusion. Instead of a clear presence of any one thing, a strong case for DV presents itself in a negative Q-tip test (one in which the pain is *not* reversed by topical anesthesia; see figure 3); negative cultures for viral, bacterial, and fungal pathogens; and a history of unremitting pain with a burning quality. Patients with these symptoms are no less anatomically alienated or verbally reticent, making the gathering of a medical history equally demanding. And though they are also likely to report having sexually "shut down," the discontinuation of sexual and genital contact does little to alleviate symptoms that are typically unprovoked. These patients can be clinically vexing and require a level of care and expertise virtually unavailable outside of specialty clinics like the VHC (Connor, Brix, and Trudeau-Hern 2013; De Andres 2015). Physical changes are not obvious to an untrained eye, and sexual despair is unappreciated by inexperienced ears. Ethnographic, anecdotal, and research-based evidence all indicate that these patients do better when seen frequently and from a multidisciplinary approach (Jensen et al. 2003; Munday et al. 2007; NIH 2012; Reed, Haefner, and Edwards 2008; Wojnarowska et al. 1997).

Since the pain of DV can be neither localized nor reversed with topical anesthesia, these women are not candidates for surgery or laser therapy. Although Zolnoun, Hartmann, and Steege's study (2003) did not include women with DV, Dr. Robichaud continued to offer nightly lidocaine ointment, counseling patients that they might still benefit from its regular application. DV patients were also referred to PT, though it does not hold the same promise as it does for women with VVS (Goetsch 2007). This is true for several reasons. First, having both tolerated and (more than likely) enjoyed vaginal penetration in their past, these patients' pelvic floors are not nearly as contracted and traumatized as women for whom genital approach *equals* pain. Second, many of these patients have had children, providing them with a slightly greater familiarity with the muscles of their pelvic floor. And third, because they are apt to seek relief sooner, their pelvic floor tension has likely accumulated over months rather than years, making its recuperation far less onerous.

The poorly understood—and clinically puzzling—nature of DV is succinctly captured by the following summary: "Some success has been reported with the use of topical anesthetics and steroids, low-dose amitriptyline and other tricyclic drugs, gabapentin, antifungal medication, dietary manipulation, biofeedback, psychotherapy, acupuncture, laser therapy and surgery" (Munday et al. 2007, 19). Treatment, in other words, is all over the place and often only moderately successful. Specialists like Drs. Erlich and Robichaud are cognizant of the need for a multidisciplinary approach but typically cannot accommodate this need in routine practice. Many of the aforementioned therapies are neither covered by insurance nor desired by patients (e.g., psychotherapy). Some are complicated by side effects, such as the disruptive sedation caused by amitriptyline and gabapentin,[7] or the likely—and poignantly counterproductive—risk of anorgasmia associated with SS(N)RIs. Effective doses of steroids carry long-term risks of immune system compromise, and, at the time of my research, the evidence regarding (often extreme) dietary modification was scant enough that VHC physicians did not feel justified in suggesting it.[8] It was challenging, to say the least, to convincingly purvey any of these therapies in solution-oriented terms.

Though they might have turned up in the clinic sooner, I found these women to be as genitally reluctant as their peers with VVS. I also found them to express equal—if not greater—degrees of ambivalence about returning to sex lives that had been defined in primarily penetrative terms. Ashley, for example, told me that she finally sought help when it "just felt like the whole inside of [her] vagina was raw" and like her partner was "wearing sandpaper for a condom." (In contrast, what Ashley told her physician was that she had "some pain with intercourse," for which she was told to use more lubricant.) She tried talking with her partner when "it got bad enough" but told me in our interview that "it wasn't really any kind of a conversation. They mostly think it's in our head—it goes along with the whole headache thing" (see Kempner 2014). Believing her physician that it was solely an issue of lubrication, however, Ashley dutifully purchased a new product; not only did this do nothing to relieve her pain, it had the unintended effect of causing her partner to believe that she was not aroused by him. When I asked Ashley if she and her partner had discussed things that he might do differently, she said, "I, um . . . I did try the lubrication. It's not anything that I'm comfortable enough to ask anything at all. You grit your teeth and hope he gets the message."

But gritting one's teeth can take a woman only so far, and the DV patients I met struggled not only to manage but to comprehend their disease process at the most basic of levels. After an emotionally draining day in the clinic, I jotted down the following field note: "These women are the ones whose sexuality we just don't want to deal with. And they themselves don't seem to want to deal with their symptoms. Or their treatment. It's messy. And it's difficult." In the end, these patients are primarily managed with antidepressants (prescribed as neuroleptic agents) (Reed et al. 2006), close clinical follow-up, and encouragement. They are also urged to seek support with the local chapter of the National Vulvodynia Association (NVA)[9] and to use online networks for additional information.

The Lichens

The final category of pain conditions is more successfully managed, and typically better understood, by the VHC clinicians. The three diagnoses that constitute this group—*lichen planus*, *lichen sclerosus*, and *lichen simplex chronicus*—are inflammatory conditions presumed to be autoimmune in nature (Byrd, Davis, and Rogers 2004; Lotery and Galask 2003).[10] The first two in particular are commonly managed by dermatologists, with topical steroids, immune system modulators, or therapeutic skin care regimens. But conventional dermatologists are not always attuned to the special needs of these patients, and several of my informants disclosed histories of inadequate and even neglectful treatment in the hands of these physicians. The VHC doctors, in contrast, attended to the well-being of these women's genitalia (including its anatomical integrity), to their sexual expectations and desires, and to the emotional fallout associated with a chronic genital condition. Control over the symptoms related to these afflictions could often be achieved easily with a combination of therapies. But a routine prescription for the nightly use of a vaginal suppository, for example, was often undermined by both the genital reluctance of the patient and the thwarted penetrative aims of a disappointed partner.

Lichen planus (LP) and lichen sclerosus (LS) are autoimmune disorders of the skin and mucous membranes. Both affect the vulva, though not exclusively: LP commonly occurs in the mouth (Edwards 1989; Smith and Haefner 2004), and LS can affect male genitalia (Friedrich 1983; Kunstfeld et al. 2003). On exam, LP is raw, moist, red, and irritated, while LS is white, dry, and leathery. Both conditions can result in irreversible labial contour changes, and LP—if not managed correctly—can lead to drastic reductions in vaginal patency (or "capacity," in clinical jargon). Both are understood in chronic terms; patients are counseled about a lifetime of medical management and that underlying cellular changes are associated with a slightly increased risk of vulvar malignancy (Renaud-Vilmer et al. 2004; Smith and Haefner 2004).

In an autoimmune disorder, the body organizes an immune response—in the form of specific cells, chemical reactions, and antibody production—despite the lack of an identifiable foreign substance or pathogen. In other words, though it is more or less normal for a body to immunologically reject an object that it does not recognize (as itself), such as a transplanted organ, this physiological response is maladaptive when deployed in the absence of a bodily threat. Autoimmune conditions, such as fibromyalgia, lupus,[11] and Crohn's disease, are life-altering not only because of the severity of their symptoms, but also because steroids are one of the most effective ways to alter an inappropriate immune response (Ahmed et al. 1999). Since steroids work by turning down the immune system, thereby mitigating the conditions' ill effects (e.g., bowel inflammation in Crohn's disease), users are rendered immunologically vulnerable while also being plagued by potentially irreversible side effects (e.g., osteoporosis, impaired glucose tolerance) (Citterio

2001; Stanbury and Graham 1998). LP and LS often baffle clinicians without vulvar expertise, whose (legitimate) concerns about steroids override their investments in the anatomical and sexual integrity of women who might lose (parts of) their genitalia without the judicious use of these medications.

The exact cause of lichen planus, according to Smith and Haefner (2004), is unknown, though "there may [be] a genetic link" (105). An LP-affected vagina acts *as if* it were responding to a bacterial or viral agent and—in addition to producing copious amounts of inflammatory discharge—becomes red, irritated, and hypersensitive. Unless halted or suppressed with steroids, the inflammatory nature of the discharge begins to permanently scar and compromise the patency and suppleness of the vagina. The vulva's contact with the discharge is also deleterious: labia can lose their suppleness, and the clitoris and clitoral hood can lose both flexibility and mobility.

The skin changes and pathology of lichen sclerosus differ from LP: a lack of vaginal discharge obviates patency concerns, but the absence of inflammation makes the changes of LS both more insidious and potentially more severe. Without generally understood markers of disease (such as redness, heat, pain, or discharge), a woman unfamiliar with her vulva may not notice her skin becoming drier, whiter, scalier, or mildly itchy; this is especially likely if she is peri- or postmenopausal, in that she may interpret these changes as part of a normal aging process. By the time her pain and itching have become intolerable, the skin changes have likely become irreversible, a diagnostic trajectory that made LS appointments hard to witness.

Loss of labial contour can be primarily cosmetic, unaccompanied by any functional impact or change. Were she to inspect, a woman might notice labia majora that were flat, and reduced in size and thickness. They might appear shiny or leathery in texture as a result of the subtle and cumulative scarring taking place. The majority of women that I have encountered—in clinical practice, personal relationships, and as informants—would neither notice nor be overly troubled by such changes, particularly if they were not associated with malignant, sensory, or functional outcomes. Vaginal scarring, however, can progress to a partial or complete stenosis (narrowing), and chronic vulvar flattening can cause the labia and clitoral hood to lose their flexibility and sensitivity. In clinical terms, the skin will have "well-demarcated, smooth whitish shiny plaques" and will be "thin and fragile with a cellophane paper-like texture" (Kunstfeld et al. 2003, 850).

When I met Mary Hudson, her LS was well controlled—she saw Dr. Robichaud at least every six months, and she joked that she had made "an altar" to the medication that was successfully keeping her symptoms at bay. During her exam, however, she described the "muffled" nature of her clitoral orgasms as Dr. Robichaud confirmed that the retractability of her clitoral hood appeared to be irreversibly impaired. In fact, Mary's clitoral tissue had severely receded several years before she found the VHC. When I interviewed her over breakfast a few weeks after we met, she gave me a fuller account of the ill-informed conditions under which her LS was detected:

> MARY: Um, I'm up in the stirrups [and all] of a sudden [the nurse practitioner] goes, "Oh my god," I mean, "OH MY GOD! Did you have a clitorectomy [*sic*]?" I'm like, "What?" . . . . She rushes off and gets the gynecologist. "We're going to do a little biopsy." . . . The good news is I don't have a, I don't have a history, a story of *suffering* in the past. I ha—, . . . and it was so, I was quickly diagnosed. Um, at that point in time I had *no* pain, *no* discomfort . . .
>
> CHRISTINE: No itching?
>
> MARY: No symptoms, no *nothing*. And Jenny[12] and I had noticed that there was kind of a whitening, and, you know, sort of what appeared to be retracting of skin and . . . And um . . . you know you just . . . because nothing, everything still w—, you know, nothing's changed in terms of *function*, um, the assumption is "Oh another sign of aging." [laughs] I didn't know this happened.

As this story reveals, the skin of an LS patient is subject to a variety of misinterpretations about the (female genital) body. Ugly, worn, aged, and strange are characteristics routinely associated with the vulva, making an LS-afflicted vulva resonant with the specter of "canned hams dropped from great heights" raised by sex columnist Dan Savage (1998, 124) and others.[13] Inured to this narrative, symptomatic women *and* inexperienced physicians often allow these changes to progress beyond the point of full recuperation; indeed, in the most extreme presentations of LP and LS, labia may need to be surgically separated in order to make the vaginal and urethral openings accessible again.[14]

Finally, there is lichen simplex chronicus (LSC), which, despite its name, is often the most easily resolved of the three conditions. Dr. Robichaud characterized LSC as the physiological outcome of an itch-scratch-itch cycle that she believed had a frequent stress/central nervous system component. Some of the earliest literature on LSC described its changes as "neurohistologic" (Cowan 1964, 562), and it is sometimes concomitant with other genital conditions. Dr. Robichaud first diagnosed Daphne with LSC and, once she had brought it under control, gained a clearer picture of her underlying VVS. In some ways, LSC represents a conceptual and clinical bridge between the lichens and the other two types of vulvar pain, in that the histologic changes can be more consistently demonstrated on a biopsy (as with the other lichens), but there is a side to LSC that is clearly psychologically aggravated—perhaps even provoked—and well sustained by cultural norms that marginalize its existence.

The initial itching in LSC is in response to some kind of stressor, which may be environmentally (a product) or internally (anxiety) derived (Foster 2002; Virgil, Bacilieri, and Corazza 2001). Whichever the cause, when a woman begins to scratch, she layers skin disruption on top of the original stressor, compounding the inflammation and skin distress and leaving the underlying problem unaddressed. Dr. Robichaud suggested to me that LSC often had an "OCD component," and she typically mined the medical history portion of these patients' visits for evidence of this disorder. I encountered only a few women with these symptoms during my fieldwork, but each time I did, Dr. Robichaud said something after the interview like "Did you hear that? She can't stop scratching." She often followed this observation with a theory about the functional nature of the scratching: a university professor's LSC, for example, "always acts up at the end of the semester when she has a lot of grading to do." On more than one occasion, I heard patients compare their LSC-related scratching with the behavior of an allergic or irritated animal: "I'm like a dog. I just want to scoot myself along the floor. It's driving me crazy!" Indeed, "driving me crazy" was a phrase frequently uttered around this disease.

Treatment of LSC is directed at eliminating the initial stressor, if it can be identified, and stopping the itch-scratch-itch cycle (Lynch 2004). This often consists of short-term steroids and histamine blockers to stop the scratching that has become a compounding stressor. Dr. Robichaud also recommended the repeated—and heavy—application of an emollient (such as Vaseline or Crisco) that could protect the skin from further injury without adding new ingredients to the local environment. VHC patients were seen frequently (once or twice a week, for several weeks) in order to support their "withdrawal" from scratching; once the skin was healed, patients might be treated with an SSRI or other psychoactive medication if the primary stressor was deemed psychological. Otherwise, the patient was counseled to avoid identified irritants and to return to the clinic promptly should her symptoms recur.

"PRETTY WHITE SKIN": RACE, GENDER, AND FEMALE GENITALIA

So who *gets* vulvar pain, and why? Prior to the publication of Harlow and Stewart's 2003 findings, the first of these questions would have been answered with a fairly homogenous demographic profile: diagnosed women were almost exclusively white/Anglo-American, educated, economically stable or insured, partnered (usually married), and heterosexual (Furlonge et al. 1991; Gordon et al. 2003; Masheb, Brondolo, and Kerns 2002; Tympanidis, Terenghi, and Dowd 2002). Aside from this profile, however, published research about the *why* of vulvar pain was relatively rare in the last decades of the twentieth century. Many dismissed vulvar pain as a psychosomatic condition experienced by a certain kind of woman, which rationalized treatment with antidepressants and other neuroleptic agents (Jantos and White 1997; Lynch 1986; Schover, Youngs, and Cannata 1992). And though some researchers paid attention to demographic factors such as race and income, studies were far likelier to discriminate between cases and controls by tracking variables such as age at menarche and first coitus, history of STIs, or comorbid conditions such as depression and fibromyalgia (Foster 1995; Harlow, Wise, and Stewart 2001; Sadownik 1999; Reed et al. 2000).

This early research was also vague on prevalence rates. But in 1990, a curious gynecologist named Martha Goetsch examined all of the patients in her general gynecological practice over a six-month period in order to establish baseline data for vestibule-specific pain. Via a combination of physical exams and interview questionnaires, Goetsch determined that 37 percent of her patients had "some degree of positive testing" and that 15 percent of them "fulfill[ed] the definition of vulvar vestibulitis" (1991, 1609). These data were significant for two reasons: first, the prospective design and large sample size ($n = 210$) helped to put vulvar pain on the clinical landscape, and, second, inclusion criteria based on both a survey and a physical exam legitimated the objective (i.e., nonpsychosomatic) nature of the condition. Using the Q-tip test described earlier (see figure 3), Goetsch established some parameters for "normal variation in sensitivity of vestibular skin" (1991, 1609) and encouraged gynecological clinicians to include VVS in their diagnostic workup of *dyspareunia*, or pain with intercourse.

Prior to this research, a lack of objective data in the literature led to inadequate clinical workups and dismissive attitudes on the parts of many providers. Nonstandardized inclusion and exclusion criteria also led to misapprehensions about exactly who was symptomatic. Collected data and published review articles from this period represented, therefore, not simply the populations who *presented* with vulvar pain but rather those who were listened to. In the late 1980s and throughout the 1990s, complaints of "pain with sex" were often labeled psychosomatic by clinicians who could not determine an obvious organic cause (Baggish and Miklos 1995; Bodden-Heidrich et al. 1999). The vulva as a primary source or site of an unmapped disease condition was barely thinkable, and women who found their way into research studies and formal clinical descriptions were those persistent enough to locate providers who were attuned to this possibility. This first wave of data, then, reflects a patient with significant resources: access to a physician, emotional resilience and persistence, adequate bodily awareness and a vocabulary to express it, and the good fortune to connect with a sympathetic provider. Unsurprisingly, the profile of these patients was that of a variably privileged white woman (Connor, Brix, and Trudeau-Horn 2013).

A slow accretion of these data led to a growing number of vulvar specialists during the late 1990s and into the new millennium. Many operated within research clinics like my field site, and some maintained more general gynecological practices. These providers now scrutinized their patients in order to articulate the clinical characteristics and criteria for vulvar pain conditions (Edwards 2003; Foster and Hasday 1997; Glazer and Rodke 2002; Masheb et al. 2000). Some of this research began to include the racial-phenotypical and socioeconomic descriptors of symptomatic women, an erratic trend that served, at first, to solidify clinical narratives about the demographics of vulvar pain. Isabelle, for example, who was one of my first informants, traveled out of state in early 2000 in order to consult with one of these early specialists after she saw him on television. During our interview, she recalled being told by one of his associates that there was a distinct—and Anglo—phenotype associated with VVS; this physician, whom Isabelle described as "Middle Eastern," even commented on her "pretty white skin" during the consultation.

Black and White Pain

During this first wave, most researchers relied on Goetsch's prevalence data and continued to assume that approximately 15 percent of the population was afflicted with vulvar pain. Indeed, the National Vulvodynia Association (NVA) strategically deployed this figure in their (successful) efforts to secure greater legislative recognition of and funding for the effects of these conditions. It was not until a phone survey study, conducted by the Harvard School of Public Health (Harlow and Stewart 2003), that Goetsch's figures were submitted to serious replication efforts. These researchers were keen to establish prevalence figures from a larger sample size ($n = 3,358$), but their interests were also informed by ongoing discussions regarding the accuracy and implications of a phenotypic profile. In other words, if vulvodynia and VVS were conditions specific to white women, it was imperative that research efforts at least include an attention to genetic predisposition and other biological markers; it was also important that associated demographic factors (e.g., income level, education) be more closely scrutinized. Finally, a correlation between vulvar pain and heterosexuality was reason to investigate the role of particular behaviors and genital practices in the development of vulvar pain.

Conscripting a larger sample size was facilitated by Harvard's survey-style approach, a method that also served to transform the analytical nature and implications of their study. A large-scale and randomly generated phone survey gave women who had not (yet) self-identified with vulvar pain an opportunity to disclose symptoms that they had perhaps not previously understood in clinical terms. Harlow and Stewart identified the need for this approach in their pilot study, noting that, when asked, "Women from the general population [were] willing to provide sensitive information on lower genital tract discomfort—a first step toward bringing notice to this understudied disorder" (Harlow, Wise, and Stewart 2001, 545).

Harlow and Stewart's interest in speaking with nonpresenting women converged with a new attention to the racial profiling of patients with vulvar pain: an unprecedented 35 percent of their sample consisted of what they called "non-white" women, a number they claimed "allowed [them] to make one of the more accurate assessments" (87) of racial distribution. And whether it was the more inclusive nature of their data, or whether other, as-yet-unidentified factors have contributed to an increased *overall* prevalence of vulvar pain, Harlow and Stewart produced results that all but contradicted the assumptions then held by clinicians. This was epitomized by their assertion that "Hispanic women were at the greatest risk of unexplained chronic vulvar pain" (2003, 85), followed closely by their finding that "there was very little difference in risk between white and African American women" (85). Although unprecedented at the time, several smaller-scale studies have subsequently demonstrated that the experience of vulvar pain is not limited to Anglo-American women (Lavy, Hynan, and Haley 2007; Reed et al. 2004; Reed et al. 2012; Ventolini and Barhan 2008). It is these studies—those that have produced new racial profiles, as well as those looking for vulvar pain *outside* the offices in which symptomatic women present—that I am characterizing as the second wave of vulvar pain research, and that I argue demonstrate the need for increased research regarding the racialized distribution of vulvar pain (Nguyen, Reese, and Harlow 2015).

Alongside this wave, several post-Harvard studies continue to suggest that vulvar pain patients are predominantly Anglotypic, although in these instances race is sometimes conflated with economic access—to a web-based survey (Gordon et al. 2003; Kaler 2006), for example, or a private gynecological practice (Arnold et al. 2006). Still other studies adduce race from biological markers, such as genetic and immunological alterations that contribute to prolonged and increased susceptibility to physiological inflammatory responses (Bachmann et al. 2006). Indeed, one group (Foster, Sazenski, and Stodgell 2004) reported that the interactions between specific genetic variations, including one involving melanin, lead to an "8-fold additive risk" of VVS for women with "light skin and red hair" (cited in Bachmann et al. 2006, 454).

What is most compelling about these assertions is that these two bodies of research—those that critically interrogate the racial makeup of their samples and those that propose a Caucasian phenotype—are not engaged in an explicit dialogue. This analytical gap is epitomized by the fact that Harlow, who coauthored the study that decisively reframed the racial distribution of vulvar pain, is also one of the authors of the "State-of-the-Art" consensus statement on vulvodynia by Bachmann et al.—a monograph that addresses race solely in terms of biological markers. In short, the relationships between race, class, and genital pain are far more complex than clinicians have thus far appreciated, leaving ample room for social scientists to, in Dr. Robichaud's words to me one afternoon, "figure it out."

My sample from the VHC was consistent with the first of these two waves: only two of eighty-two women from whom I collected demographic data defined themselves as Latina, and one defined herself as South Asian; none were African American, and the remainder identified as white/Anglo/Euro-American. Questions about the whiteness of these patients typically turn on the issue of access, specifically the racially inflected continuum along which health care is distributed in the United States (Dressler, Oths, and Gravlee 2005; Krieger 2005). (This is distinct from my discussion in the previous chapter, where I define "access" in terms of the emotional and linguistic reticence compromising women's ability to obtain dedicated vulvar care). Within this frame, it is reasonable to assume that women of color—who are often poorer in the aggregate—lack access to the type of specialty care necessary for the proper management of vulvar disease conditions. And, in some ways, this is the case at Riverview: the clinic is a locus of rare expertise that is unevenly available to low-income and other resource-poor women. But this is only part of the story.

Like most teaching hospitals, Riverview has a GYN resident clinic where low-income and publicly insured women receive subsidized or free treatment from physicians-in-training (but not without long waits and a good deal of bureaucratic surveillance). But the resident clinic generates some noteworthy contradictions, making Riverview an especially interesting site from which to reexamine the issue of access. Because the hospital houses one of the few specialty vulva clinics in the United States, any symptomatic woman who finds her way to the hospital's provider pool is at least one step closer to accessing the VHC's expertise. In theory, this should be made even easier by the fact that the clinic's two physicians work with, teach, and supervise the hospital's gynecology residents in multiple areas of the hospital, including the resident and vulvar clinics. In short, and unlike many other hospitals and clinical practices, the services received by lower-income—often African American and Latina—women in the resident clinic at Riverview are typically provided by the same doctors and residents caring for white patients in the vulva clinic.

This configuration renders questions of health care access—the frame through which many disparity scholars (Smedley, Stith, and Nelson 2003; Smith et al. 2007) explain the demographics of vulvar pain—far more complex. As Anne Pollock has recently demonstrated in her work on race and heart disease, "Medicine both represents *and intervenes on* racialized bodies" (2012, 3; my emphasis); as a bio-ontological category, race "travels" between providers' offices, resident clinics, and research protocols in ways that are flexible, plural, and "noninnocent" (5). Indeed, the profile of vulvar pain at Riverview is rendered even more complex by a set of assumptions around which many physicians practice, a "durable preoccupation" (8) with race that transcends issues of medical availability and relative expertise. This was made clear to me one morning when a resident nonchalantly remarked that "vulvar pain is *white* and pelvic pain is *black*" (my emphasis).

To be sure, assumptions about race and disease adhere to far more disease conditions than vulvar pain (Anderson 2003; Krieger 2005; Tapper 1999). That said, it is important to at least briefly unpack the racialized "story" (Pollock 2012) being told by this resident, in order to more carefully think through the relative roles played by structural, ideological, biological, and sociocultural factors in the demographic distribution of vulvar pain conditions. By locating her remark within a medical imaginary of human difference, where race and disease map neatly onto one another, rather than within a body of clinical literature (that only weakly supports her assertion), the resident demonstrates "how racial inequality becomes embodied—literally" (Gravlee 2009, 47). Like the eighteenth- and nineteenth-century physicians described by Tapper (1999), whose fixation on the "blackness" of sickle cell disease compromised their ability to understand its distribution across a wide range of phenotypes, this physician is confounded by the presence of vestibular pain in the body of a dark-skinned woman. Kempner (2014), in her discussion of persistent gendered imaginaries in the diagnosis of migraine, suggests that "cultural stereotypes may be more resilient than the emergence of new biological knowledge" (157) and that "medical researchers engage in . . . dangerous form[s] of . . . essentialism" (151) when they fail to interrogate the sources of their disease narratives. These kinds of contradictions are one more reason for social science and humanities scholars, for whom the "recordable" (Pollock 2012, 85) categories of race and gender are assumed to be complex and polyvalent, and for whom these categories represent entangled forms of biocultural citizenship rather than discrete and additive demographic realities, should continue to study vulvar pain conditions. While biologically inflected, race and gender are also "deeply embedded in sociocultural systems" (Gravlee 2009, 54); as such, they are best analyzed by accounts that take the mutually influential relationship of these two realms into account.

Moreover, the confusion provoked by these overlapping statistical and anecdotal realities is compounded by an unevenly demonstrated empirical relationship between race and chronic pelvic pain (CPP), the "black" pain in the resident's remark (Apte et al. 2011; Jamieson and Steege 1996). CPP is defined by Haggerty et al. (2005) as "menstrual or nonmenstrual pain of at least 6 months' duration that occurs below the umbilicus and is severe enough to cause functional disability or require treatment" (293). It is often described as a diagnosis of exclusion because of the high likelihood (up to 60 percent) that a clinical workup will fail to reveal specific pathology (Apte et al. 2011; Nelson et al. 2011). Because it can arise from a number of bodily systems, managing CPP can be burdensome (Cacchioni 2007; Cacchioni and Wolkowitz 2011; Jarrell et al. 2005); it is less characterized by "that one spot" than it is by generalized pain sensations across the pelvic cavity and viscera. Whereas the coital pain associated with vulvar pain is (primarily) superficial and provoked by penetration, CPP is deeper and aggravated by thrusting. In contrast to vulvar pain's "burning" sensations, CPP is often described as aching or throbbing. CPP also has an older medical "reality" than does unexplained vulvar pain, and the racialized and class-based discourses through which it is understood precede the establishment of specialty vulvar clinics (Learman 2005).

And though women with CPP also restrict their sexual activity, the dull and more imprecise nature of their symptoms allows them to engage in a slightly more expansive behavioral repertoire. This pattern is related, of course, to partner expectation, varying levels of sexual coercion, and widespread uncertainty on the part of many heterosexual couples regarding the normality of sexual pain. But the question raised by the resident has little to do with these distinctions or with which so-called racial groups are statistically afflicted with either type of pain. Rather, it reveals the degrees to which pelvic and vulvar pain patients are likely racialized (Omi and Winant 1994) through their symptoms and and raises the question of whether the *vulvar* pain of an African American woman can be apprehended as such if she is first encountered in the resident clinic—the space of *pelvic* pain (Ahmed 2007).

Hysteria, White Femininity, and the Intersectionality of Pain

Finally, it is worth noting that the majority of women diagnosed with vulvar pain bear an uncanny resemblance to many of the women who suffered both the symptoms and the "cures" of neurotic hysteria at the turn of the twentieth century. By critically juxtaposing these two groups of patients, I invoke one element of vulvar pain about which many feminists have previously written: the elite marginalization of a group of educated and economically privileged white women in the hands of gynecological medicine.[15] Historical analyses of hysteria allow us to contextualize contemporary medical narratives about the neuropathic nature of vulvar pain and deepen an investigation of culture and physiology, particularly as these pertain to the intersection of class, race, and gender. In this context, I suggest that vulvodynia and VVS, though real physical conditions, are circumscribed and informed by what Scheurich calls a "symptom pool"—a historically and culturally specific social structure "through which distress is experienced and expressed" (2000, 461).[16]

The feminist paradox of hysteria is that its diagnosis relied on both the subordination *and* near hypercivilization of (some) white women. Routinely conflated with children, criminals and deviants, animals, and nonwhite men, European and US women at the turn of the twentieth century occupied the savage half of the civilized/savage spectrum (Ehrenreich and English 1973, 1978; Haraway 1989; Schiebinger 1993), and women who sought an education or intellectual occupation were resisted by dominant and inferiorizing patriarchal discourses and structures. These pursuits were also unavailable to nonwhite and laboring-class women, but for distinct reasons. Discourses of racial superiority depended on clearly drawn and biologically rationalized divisions between alleged races, and sex-linked inferiority depended on many of the same rationales (of white male supremacy, for example) (Briggs 2000). In this period of increasing cultural contact, however, the salience of racial difference took social precedence, and white women were granted partial inclusion in an allegedly superior Caucasian race. This meant proscriptions against educational and occupational advancement needed to be carefully couched, as the labeling of white women as too coarse, savage, or deviant threatened a developing and race-based social hierarchy (Horn 2003; Schiebinger 1993; see also Kempner 2014, 28–33).

The emerging science of neurology provided one of the tools with which women could be both included in a superior race and simultaneously excluded from its full spectrum of benefits. Delicate nervous systems were mobilized in order not only to rationalize the exclusion of these otherwise privileged women from intellectual pursuits but also to demonstrate their very real—and biological—differences from nonwhite women (Briggs 2000; de Marneffe 1996; Jackson 1987). Preserving the purity and refined nature of this physiology was the basis for the physically confining (and socially isolating) "rest cure" developed by neurologist S. Weir Mitchell and propagated by many of his colleagues. Justification for this cure was buttressed by the astounding variety and mutability of concomitant physical symptoms manifested by these women, including choking sensations, convulsions, gastrointestinal problems, sleeping and breathing disorders, and depression (Wilson 2004; Wood 1973). The biopsychosocial nature of hysteria makes credible Elaine Showalter's assertion that "hysteria is a mimetic disorder; it mimics culturally permissible expressions of distress" (cited in Scheurich 2000, 462): hysterics who were denied social experiences such as education acted out both the roles through which they were interpellated (swooning and fainting) as well as their refusal to "swallow" or comply with the limitations imposed upon them (choking and convulsions) (Chisholm 1994; Cixous and Clément 1986; Merleau-Ponty 1962).

In a thorough account of fin de siècle European forensic anthropology (and its role in the discourses of deviance that informed an emerging eugenics movement), David Horn (2003) describes how the science of algometry (pain measurement) serviced the social hierarchy that informed the experience of conditions like hysteria. A heightened sensitivity to pain was equated with a more discriminating and civilized sensibility, and algometry was practiced by prison physicians and criminologists in order to assign individuals biologically derived—and therefore immutable—places along the social ladder: "Pain was not enlisted to extract the truth about illegal *acts*, as had been the case with torture, but rather to produce evidence of the biological nature of individuals and groups, and about the *dangers* that accompanied a 'failure to evolve'" (90; emphasis in original).

Within this scientific discourse, pain perception was not only proportional to one's level of civilization, it was also linked with one's moral sensibilities. Upper-class women with "pretty white skin" remained paradoxically positioned—inferior to and coarser than men, but more civilized than nonwhite or working-class women (Schiebinger 1993; Wood 1973). Elaborate narratives were constructed to support this multipositioned social location, so that (white) women were variously defined as: less sensitive to touch, but more sensitive to pain; more likely to fear pain, and therefore complain of it earlier; or more likely to be "irritable" than men, irritability "being the 'incipient, brute form' of sensibility" (Horn 2003, 99). The bodies of allegedly less evolved women—prostitutes, laborers, women of color, and immigrants—were not capable of these discriminatory perceptions; this meant not only that they could remain at the social margins, but that they would not experience the symptoms and diseases specific to white upper-class women.

In this historical context, I am struck by the repeated—and increasingly specific—descriptions of the "damaged" and hypersensitive nerve fibers of women with vulvar pain, particularly when these "nervous system" anomalies are linked with a white phenotype. Clinical descriptions of vulvar pain include "regional heightened [skin] responses" (Foster et al. 2007, 346), "exquisite [vestibular] tenderness" (Bergeron, Binik et al. 1997, 27; Leclair et al. 2007, 53), "increased innervation" (Tympanidis et al. 2002, 1021), and "exaggerated inflammatory reaction" (Bachmann et al. 2006, 454); these descriptions dictate treatment strategies that are couched—to both patients and clinicians—in terms of "calming" the fired-up nerves that precipitate these conditions (Foster, Dworkin, and Wood 2005). Women with vulvodynia, whose pain cannot be behaviorally or anatomically circumscribed, are particularly susceptible to these narratives: suffering from generalized, poorly understood, and "uncontainable" (read: excessive) symptoms, these patients are almost always offered some form of neuroleptic medication. Ironically, and uncannily resonant, the drugs that most effectively reduce a woman's pain perception have a sedating effect that can preclude intellectual and functional activity (De Andres et al. 2015).[17] Once attuned to this insight, it became difficult for me to witness the diagnostic process—whereby the application of a topical anesthetic transforms a writhing and contorted woman into an uncomplaining and compliant patient—outside of a "hysterical" framework.[18]

As in hysteria, vulvar pain's symbolic meanings cannot be extricated from their biological reality. In Scheurich's (2000) terms, VVS and vulvodynia do not constitute "discrete syndrome[s]" (465) but exist, rather, as "the form that illness necessarily takes in a complex, social, symbolic creature such as a human being" (465). In chapter 5, I argue that an inability to participate in penetrative intercourse indexes, at least partially, a "postfeminist" (Gerhard 2005; Johnson 2002) ambivalence toward the forms of (hetero)sexual expression available to these relatively privileged and mainstream women. Here I suggest that genital "burning" that represents pain and refusal—rather than arousal and desire—functions mimetically, providing patients with the space to more deeply explore the nature of their sexuality while they "sit out" the coital activities expected of them by their husbands, friends, and the culture at large (Jackson 2006; Potts 2002).

The resurrection of hysteria offers feminists and critical theorists of the body something more than another iteration of a gendered somatoform disorder. Indeed, I suggest that if "hysteria, by whatever name, is alive and well" (Scheurich 2000, 462), then we have an opportunity to more effectively analyze its relationship not only to gender but to race and class as well. This might mean thinking through vulvar pain diagnoses as constitutive, rather than reflective, of a privileged whiteness. This move allows us to more critically examine claims of disparate access that, while possibly accurate, do not fully explain why vulvodynia and VVS either erupt primarily in the bodies of privileged (white) women or why they are only medically evaluated when they do. In the following section, I close with an ethnographic description of the clinic itself in order to flesh out the space and the people through which this attention is bestowed.

THE CLINIC

More virtual than brick-and-mortar, the VHC as a "clinic" operates within a sprawling and ultramodern research hospital and medical school (Riverview) located in the northwest United States. The VHC was founded in the early 1990s by an obstetrician/gynecologist (OB/GYN) named Dr. Jenkins, who, though he continued to serve as its director, saw only long-established patients during the time of my fieldwork; Drs. Robichaud and Erlich, who joined the clinic shortly thereafter, saw all new patients while I was there. The VHC was located within the hospital's larger Women's Health Center (WHC), which has been nationally recognized for its expertise in clinical care, research, community outreach, professional education, and leadership,[19] meaning that the clinic is situated—materially and discursively—within a competitive and well-funded biomedical research institution and enjoys the privileges and prestige consonant with that association. On a slick and colorful homepage, for example, "Vulvar Health" is listed as just one of sixteen highlighted specialties offered by the WHC (others include midwifery, integrative medicine, and urogynecology). Potential health care consumers are promised an "unmatched" level of care for the "unique challenges" posed by (their) vulvar disease.

But in other important ways, the relationship of the VHC to the WHC is more like that of an annexed temp worker, an affiliation made most evident by its long-term absence from the hospital's main website[20] and the small number of appointments it was allotted each month. This constricted support translated into the one half-day per week that half of the WHC's physical facilities, and the schedules of the two expert staff physicians, were dedicated to patients with vulvar pain: on Thursday mornings, aka "vulva clinic," the doctors were supported by medical assistants, nurses, and front-desk staff who worked full time in the WHC and were also specially trained to understand and respond to the particular needs of women who presented with this cluster of symptoms. Because the VHC operated as a referral clinic, almost all of its patients came through an outside physician or other health care provider. Particularly savvy and persistent patients occasionally secured their own appointments, but a referral model helped to ensure that patients who suffered from other conditions did not crowd the few slots that a short half-day per week allowed.

To secure one of these appointments, a woman (or her provider) needed to report *some* version of symptoms that sounded like chronic and unexplained vulvar pain; they needed to, in Mol's words, "enact" (2002, 40–41) the condition around which the clinic was organized. Schedulers were therefore trained to screen out callers whose complaints did not require the expertise of Drs. Robichaud and Erlich. Such patients might have been seen by either of these two doctors, perhaps even in the same physical space, but unless there was a high suspicion that their symptoms were vulvar in nature, they were not given the extended initial appointments that characterized Thursday mornings. It was not unusual for an initial vulvar appointment to last one and a half hours, not including administrative time, and it was common for new patients to wait three to six months for a consultation. The fact that vulvar visits were significantly longer than standard gynecology appointments was logistically significant: many of these women had been symptomatic (and possibly inadequately treated) for a number of years, and it took a fair amount of time for clinic staff to sort through and collect the pieces of these medical histories that were immediately pertinent and amenable to intervention. Since the majority of patients reached the clinic through another provider, their medical histories were often represented by hefty stacks of diagnostic and treatment notes. These were routinely faxed over from at least one medical office, often incomplete, and rarely given the kind of attention that patients believed they deserved.

Thursday appointments were also special for a set of less quantifiable reasons. The shame, reluctance, and aversion that many women feel toward their genital bodies compound a set of symptoms that remain a physiological puzzle for most clinicians. Genital distaste is both individual and collective—evident in the bodily and the linguistic practices of women with symptoms, some health care providers, partners, insurance companies, and the culture at large. Women who staff the vulvar clinic are acutely aware of the enormous personal barriers that patients must overcome in order to relate their story and ready themselves for treatment strategies that they anticipate will be uncomfortable—physically, emotionally, and socially.[21] Indeed, explaining some of these treatment options (PT, surgery, at-home topical anesthesia) was time-consuming in itself, made more so by the linguistic reticence and affective ill preparedness of many patients at their initial visit. Clinic physicians and staff, therefore, needed to care for Thursday morning patients in a manner that simultaneously recognized and attempted to transcend these very personal obstacles.

Drs. Robichaud and Erlich functioned as a team who shared a common purpose of keeping vulvar pain on the clinical and institutional map. The first patients were typically scheduled at 8:00 a.m. (Dr. Robichaud, who took one day off per week to spend with her family, would occasionally squeeze someone in as early as 7:00.) The doctors arrived anywhere from thirty to sixty minutes early in order to catch up on the myriad tasks that constituted indirect patient care. While the doctors settled in, the medical assistants (Gia, Katie, and Leah) and vulvar nurse

(Jane) moved around busily: setting up exam rooms, readying charts, and retrieving the first patients from the waiting room. Patients were weighed on a scale in the hallway and, once "roomed," had their vital signs taken (blood pressure, pulse, respiratory rate, and temperature) and were asked to provide a brief description of why they had come to the clinic. New patients remained clothed until they met their physician and were allowed to sit on a built-in, padded bench at one end of the room. Returning patients, on the other hand, got undressed from the waist down and awaited their doctor while sitting on the exam table, covered with a crepe paper drape sheet.

The physical space of the clinic was laid out as one long hallway with four exam rooms off its left side and two exam rooms, a bathroom, and a utility room off its right. The front end of the hallway housed a cramped and hyperutilized "pod": a tiny and doorless cubby that sat three to four, semi-comfortably, at its wraparound desktop and could, when needed, hold two to three more adult bodies at one time. The desktop held three bulky computer monitors, and the room housed an eclectic collection of rolling office chairs; these were constantly moved in and out of exam rooms to accommodate partners, medical students, residents, and, in my case, an anthropologist. The pod was where the doctors started (and ended) their days, where clinical consultations and instruction were often carried out, where lunches were frequently (and incompletely) eaten, and where the majority of vulva-related paperwork (e.g., fact sheets and referral forms), formal literature (e.g., the *Atlas of Vulvar Disease*), and instructional materials (such as the demonstration vaginal dilator kit; see figure 5) were kept.

Thursday mornings were, in a word, exhausting. They were also, in many words, touching, efficient, gloomy, exhilarating, devastating, puzzling, hopeful, and hilarious. They were filled with kindness, gratitude, and expertise. Women who waited up to six months to have pain of anywhere from one to ten years' duration carry a particular blend of optimism, inertia, and despair. The fact that the pain is of a sexual nature can add a conflicted charge to their affect, but this charge is muffled by the bodily shame that engulfs their genital experience. The doctors at the VHC had purposefully chosen to specialize in vulvar pain, yet they constantly risked being overwhelmed by the fraught mix of needs, desires, and emotional limits expressed by their patients.

The fact that these delicate negotiations could not be adequately conducted in a standard fifteen-minute appointment indexes the limited resources that gynecological medicine can contribute to novel—and vulvar-based—sexual imaginaries. One night I had dinner with Dr. Robichaud and Jill, the clinic's sex therapist; over sushi and sake, we talked in earnest about our respective reasons for being involved with the VHC. Dr. Robichaud spoke clearly, telling us that her vulvar work was the most gratifying part of her practice, primarily because it allowed her to more fully engage with her avocational interest in sexuality. I immediately began to clarify her statement by suggesting that *all* of an OB/GYN practice necessarily involved an engagement with female sexuality, but, even as the words began to come out of my mouth, I realized that I was wrong.

As we conversed, I remained astounded that I had never before appreciated the depth of this reality, even (especially!) while I practiced. As I listed the kinds of care most frequently provided by OB/GYN practitioners—contraception; menstruation and menopause management; pregnancy and childbirth; the prevention, diagnosis, and treatment of sexually transmitted diseases (STIs); breast health; and genital cancer prevention—we began to wince with recognition and resignation. It was unsettlingly easy to recall all of the ways that these health care problems are efficiently handled without any explicit attention to, or frank discussions of, sexuality. Indeed, modern-day contraception epitomizes the abstracted sexual body, in that the most heavily prescribed methods (hormonal pills, patches, injections, and IUDs)[22] absent women from their genitals in ways that barrier methods cannot. While diaphragms, caps, and condoms necessitate direct—and manual—contact with labia, cervixes, and penises in order to be used correctly, the methods chosen by most women and couples in the contemporary United States who are hoping to avoid an unwanted

FIGURE 5. A full set of vaginal dilators

Image used with permission of Jane Silverstein.

pregnancy allow them to efficiently bypass these body parts in order to secure that outcome.

The tension between this traditional mode of gynecology and the attention to female sexuality on which adequate treatment of chronic genital pain depended surfaced in waves in the vulvar clinic. Symptomatic women, along with their doctors and partners, awkwardly (and inconsistently) generated *and* suppressed novel behaviors and language practices through which new heterosexual orientations could emerge. This is partly because marginalizing the vulva allows all of us, patients in particular, to continue to invest in the correctness, naturalness, and even sacredness of penile-vaginal complementarity. A dearth of materials and discourses rooted in alternative sexual imaginaries confounds conventional gynecologists' efforts to transcend this narrative. But a vulva that existed *for itself*, a vulva whose primacy was enacted through the clinic's focused efforts and apparatus (Mol 2002), was made available by the VHC, though not all patients were equally able to identify or grasp its sometimes fleeting presence.

The Staff and Their Roles

The person most explicitly responsible for helping patients craft these alternative imaginaries was Jill, a therapist who specialized in sexuality and grief work (postpartum depression, specifically). In the clinic's earlier days, Jill evaluated all new vulvar patients, but these automatic consultations ceased when the hospital stopped subsidizing her time. Because sexual counseling has been shown to play a role in the reduction of vulvar pain (Bergeron et al. 2001, 2008; Connor, Robinson, and Wieling 2008), Jill's relative availability to clinic patients merits deeper consideration, as the majority of diagnosed women do not pursue it. In my thirteen months in the clinic, and for reasons both emotional and economic, fewer than five of the eighty-two patients with whom I interacted sought the help of Jill or another sex therapist.

This low percentage was partly due to the fact that patients interpreted their diagnoses as physical rather than psychological. Moreover, many of the women I met repeatedly stressed that their relationships were "not about sex . . . thank God!" By the time they reached the clinic, patients' relationships were in various states of upheaval; the majority, whose pain had been present for an average of three to five years, had adopted coping mechanisms and behavioral strategies that helped to preserve marital or relationship harmony. These diagnostic delays were one reason that some women dragged their feet about sex therapy: their emotions were not nearly the "burning" issue that their painful skin was, and many failed to appreciate the difference that "talking about it" would make in their lives. The fact that sexual counseling cost upward of $150 per session, and was typically not covered by insurance, made sex therapy even less appealing. Rather, patients saw the pharmaceutical, surgical, and other medical management of their symptoms as the path to future participation in "normal" sex.

Unlike Jill, Jane—aka the "vulvar nurse"—had no avocational interest in sexuality but had a solid role in the clinic, not only on Thursday mornings but any time that patients needed advice, support, or triage. Jane was both dry-humored and no-nonsense and had come to the WHC and vulvar clinic less than a year before I began my fieldwork. Because vulvar pain and disease were relatively invisible outside the hospital, Jane spent a lot of her time on the phone, educating and coordinating with a variety of invested parties: physicians and other health care providers, pharmacists, insurance companies, physical therapists, partners, and patients themselves. And because one clinical consultation, even one that lasts an hour and a half, is an impossibly inadequate response to the years of suffering that have preceded it, much of Jane's time involved the simple and assuring translation and reiteration of the physicians' often intricate plans of care.

Dr. Robichaud once described Jane as a "bulldog" to one of her patients, assuring her that Jane would fight to secure insurance coverage for an off-label medication. While not exactly an antidote to the tenderness and emotional intensity that could easily dominate clinic interactions, Jane's solution-oriented approach was critical in meeting the practical needs of patients once they had exited the clinic's cocoon. In many ways, Jane embodied the normalizing, even neutralizing, work of the clinic: she came from an unrelated medical specialty, picking up vulvar work as little more than a novel set of nursing skills, and learned to cultivate a professional detachment from the nonclinical stories that clung like Velcro to every facet of these conditions. Symptomatic women needed Jane to remind them that their pain, like any other clinical condition, could be confronted outside the amplified and culturally charged discourses surrounding its source.

In the VHC hierarchy, Jane's skills and developing expertise were located below those of the physicians and above those of the medical assistants (MAs). The MAs rarely ran out of tasks: putting charts together, retrieving and organizing faxed records, cleaning and preparing exam rooms, greeting patients and gleaning the basic elements of their history, and filling out lab and requisition paperwork associated with diagnostic testing. At the time of my fieldwork, two of them—Katie and Leah—were in their early to mid-twenties and were working at the hospital in order to inform their future career paths. Gia, in contrast, was a married grandmother who had been with the clinic for many years and planned to remain there until she retired. MAs were usually the first to hear the troubling stories that women brought to the clinic. And though they were frequently moved by these stories, their multiple responsibilities did not allow them to linger too long with any patient.

Depending on the day, the second set of ears to hear these stories frequently belonged to an OB/GYN resident, a medical school graduate who had completed an internship of one to two years and was now specializing in obstetrics and gynecology. During my fieldwork, the vulva clinic was an elective rotation for the hospital's residents, and most of those that cycled through had worked with one or both of the clinic's physicians in other capacities. A new resident arrived at the start of the month and typically rotated between Drs. Erlich and Robichaud on alternating Thursdays until they were replaced. Previously trained, these physicians were more than capable of gathering medical histories and making preliminary assessments. They would then present these to the doctor with whom they were working that morning, and—depending on the day, the doctor, and the pace of the clinic—a pedagogical exercise might ensue, during which the resident was asked to elaborate their clinical impressions. Busier days allowed for merely a brief summary, followed by both women going into the exam room together and the doctor using a quick patient interview to fill in what she believed to be missing from the resident's account.

On a routine morning, then, the hallway, exam rooms, and pod were peopled with six to eight professional bodies through whom the vulva clinic effectively "surfaced" (Taylor 2005). Each of them, as well as the front desk staff among whom patients sat in the waiting room, absorbed some of the hopeful or resigned energy that had crystallized for so many women in the years leading up to that day. Their professional presence, in other words, the stability and reality of their job in a vulva clinic, helped to normalize bodily experiences that had, until then, existed far "outside the range" (Brown 1995, 100) of what many of these patients had been taught to (sexually) expect. On every other level, in every other area of their lives, vulvar pain was an extraordinary event, one that challenged the credulity of the husbands and doctors of symptomatic women, people who could see nothing wrong with their complaining bodies. The clinic, in contrast, was brimming with the conviction that these symptoms were not only real but that they were amenable to routine medical interventions.

3

ACCUMULATION

The Materiality of Absence

———•·•———

Louise's presence in the waiting room was impossible to miss. She was middle-aged, white, and well put together; very petite, almost diminutive, she wore gray yoga pants and a pink cotton tee shirt covered with a quilted pink hoodie that she left partially unzipped. She kept her graying and curly hair loose and chin length, and she wore neither makeup nor jewelry to the clinic. Louise was accompanied by her husband, Niko, a tall and solicitous man who appeared to be in his late fifties or early sixties. Niko's appearance and demeanor were striking—handsome and olive complected, he was very well dressed in a dark suit, and he spoke with a vaguely European accent. But it was neither Louise's well-kept appearance nor the presence of Niko that drew uncomfortable and curious glances from the staff and patients in the waiting room. Rather, it was Louise's body that provoked unrest in the bodies around her, including my own. What had us antsy—squirming if we were seated, and more than a little off-guard if we were just passing through—was the way that Louise inhabited her chair. That is, she wasn't sitting in it at all.

New patients in the vulvar clinic were asked to complete a substantial number of questionnaires before they were brought back to an exam room. So detailed were these forms, completing them could take up to a half an hour, depending on the extent of a woman's history. As she described it to us later, Louise came to the clinic that day feeling as if she were "sitting in fire," a symptom she'd had for four months. Because of this, she decided to kneel on the floor in front of her chair and rest her clipboard on its seat, rather than subject her pelvic floor and vulva to the "stabbing pains" that were aggravated by contact with external stimuli. In the contest between immediate self-care and waiting room decorum, Louise's pain tipped the scales, and she worried little about the nervous energy that vectored around the room in response.

As it turned out, Louise's story was an anomalous one, in both the intensity and the relatively recent onset of her pain: most clinic patients counted their pain in years rather than months and, though profoundly disruptive, most had also made behavioral and lifestyle adjustments that made it more tolerable. But, as she told me on the day that we met, Louise's distress and desperation were so great that she had been looking at euthanasia websites when her son called to tell her about the vulvar clinic. She continued that she didn't know what she would have done had she been forced to wait the three to six months that most patients did (she had secured an unusually expeditious appointment because of a cancellation). Even the normally staid Dr. Erlich was mildly overwhelmed: when we stepped outside the room to allow Louise to undress, she said to me, "These are the ones who make you think, 'Oh God, I hope I know enough to help her!'"

It was this sense of urgency—performed on her knees in the waiting room (and eventually on the exam table), and then eloquently narrated during her two hours in the clinic—that shaped the framework through which Louise ultimately interpreted her symptoms. Already living with several autoimmune conditions[1] (fibromyalgia, interstitial cystitis, and irritable bowel syndrome), Louise had made major changes in her life, including the removal of as many environmental irritants and toxins as she had the resources to do. She also complained of her inability to wear anything but loose-fitting clothing, and, not surprisingly, she could not tolerate sexual intercourse. Louise linked the onset of her afflictions to a year in which she had grieved the deaths of multiple loved ones and, within this context, came to apprehend her vulvar pain in terms of the losses she was sustaining. While we sat together in the exam room, waiting for Dr. Erlich to return with a prescription, Louise rolled on to her side to take the pressure off her genitals and tried to describe the state she had been in prior to her appointment, the one that had led her to search for assisted suicide websites:[2] "I asked the universe. I had to ask—'What *else*? What *else* can you take away from me?'"

COLLECTING

In addition to the absences with which we often associate loss, the recent vacancies in Louise's life can also be viewed as accumulations: sedimentary layers in the experience of vulvar disease. Women with genital pain move through their symptoms and medical consultations in a state of haphazard collecting, gleaning a set of accoutrements specific to their lived experience: diagnoses (mostly inaccurate), prescriptions (usually ineffective), advice (often erroneous), marital discord (mostly reparable), and an increasing and inevitable sense of dread that their pain will go on forever. These burdens articulate with broader social processes that inform US women's personal and collective relationships with their genital and sexual bodies; in short, the notion of genital pain as loss is complicated by the number of women—with and without pain—that have never "had" their vulvas to begin with. The lexical and visual absence, disparagement, and inconsequence of female genitalia are so commonsensical that even women with debilitating pain will lack the words to describe its location. I suggest that this loss for words be understood as a deposit—a stratified layer of socially enforced silence that is an integral component of coming to terms with genital pain.

My notion of collecting draws on Mary Weismantel's (2001) account of racial accumulation in Andean South America. Weismantel is neither the first nor the only scholar to elaborate race's corporeality (Gravlee 2009; M'Charek 2013; Reardon 2012); moreover, and following both Young (2005) and Butler

(1993), a wealth of feminist scholarship has demonstrated the thoroughly embodied nature of gender (Fausto-Sterling 2000; Springer, Stellman, and Jordan-Young 2012; Wilson 2004). I turn to Weismantel's *Cholas and Pishtacos*, however, because of its ethnographic attention to the ways that bodies *acquire* biology through daily practices, material possessions, and social exchanges. The feet of a typical gringo in the late twentieth-century Andes, for example, are as materially inflected with economic privilege—as marked by their access to well-fitting shoes and mechanical transportation—as the (often) impoverished feet of an indigenous person are by their lack of the same. In this account of racialized embodiment, bodies "assume" (Salamon 2010) peculiar and distinct shapes through their participation in and exposure to hierarchically organized activities, via social categories that include race, gender, and social class. We can say that these categories aren't real only by ignoring the all too real divisions of labor, resources, power, and pleasures that are disproportionately accumulated by different kinds of bodies in historically and geographically specific ways. Material repercussions that are unevenly distributed along racial, classed, and gendered lines need not be genetically encoded nor inheritable for them to be biologically salient. Bodies change, sometimes permanently, due to what they accumulate along the way. Genitalia are no exception.

Disparaging discourses are collected by the genital bodies of US women, settling and shifting as embarrassing and alienated sediment, a process made evident during my informants' first verbal exchanges with the clinic physicians. The simple prompt "Tell me why you're here" was enveloped by a profound sense of heaviness: tongues thick with awkward residue, shoulders stooped, eyes downwardly cast, and words so leaden they felt like they needed to be picked up off the floor. My hope is that, in putting forth the detailed transcripts of these encounters, readers can experience some of the feelings associated with the dense fog of vulvar pain—the lack of accurate information, the linguistic confusion, the hysterically based thinking, and the hints of skepticism and disbelief that inform their symptoms. I hope to demonstrate the palpable sense of opacity to which I bore witness over my thirteen months of fieldwork: a gendered miasma of unanswered questions, impossible (hetero)sexualities, and pathologically inscrutable female bodies.

Bodily accumulation is a process, both random and ordered, and mediated through cultural norms, expectations, and structures. Some accumulations are obvious, even physically graspable, like medical records from previous providers that are faxed to the clinic in literal reams. Others, such as the layers of *genital alienation* and *unwanted genital experience* that are stubbornly lodged in corporeal sediment, have a far more nuanced nature and precipitate even heavier burdens. In the next section, I use my background as a feminist medical clinician to further examine these layers. Evaluating the relationship of my own body to those of my patients, I narrate a professional and personal negotiation with genital baggage that can be both discarded and collected. Following this analysis, I use field notes and patient narratives to capture the range of physical and emotional effects that are accumulated in the months and years leading up to an appointment.

I theorize accumulation as dynamic layering—where idiosyncratic configurations of personal resources and socially structured realities constantly shift, settle, and erode in individual bodies that move in and out of relief-seeking behavior. In the last section of the chapter, I introduce the diagnostic role of physical therapy (PT) by narrating the tension and "holding pattern" discovered in my own pelvic floor during a biofeedback session with an informant. I use this physiological "fact" to describe the diffuse and deep-tissue discourses upon and through which the more acute and eruptive sensations of vulvar pain are lived.

Alternative Baggage

> Spend a week telling people you know and meet that you are working on a book about pelvic exams. If your experience is anything like mine, your statement will be met with a variety of reactions: nervous laughter,

surprise, horror, blank stares, suggestive winks, embarrassment, anger, excitement, disgust, discomfort, absolute silence. But silence is rare, at least from women. (Kapsalis 1997, 3)

It has long been clear to me that the topic of female genital discourse incites particular kinds of trouble. Much of my research for this book focused on acknowledging (and sometimes provoking) that trouble, delineating its cultural provenance, and examining its consequences. The explicitly genital nature of my questions made scholarship arduous at times, and I was often hard pressed to locate texts that offered both theory *and* content regarding the cultural aspects of the vulva. When I discovered Terry Kapsalis's (1997) *Public Privates: Performing Gynecology from Both Ends of the Speculum*, it became an important exception.

As with the "inherent problem of the female pelvic exam," my work raised eyebrows both individual and collective because, as she argues, it "*necessitate*[d] the *public* exposure of the shameful female *privates*" (Kapsalis 1997, 5; emphasis in original). Kapsalis astutely measures the socially transgressive nature of this process and locates at least one source of the trouble that she and I provoke, along with the discomfort we subsequently share. Interrogating her own project through the lens of mainstream discourses, Kapsalis asks us to consider "what *kind* of woman writes a book about pelvic exams?" (8). By framing her research interests in this way, Kapsalis takes cues not only from the professional and personal interlocutors described in the paragraph just cited but also from conventional and hegemonic medicine. She contextualizes her critical reflexivity by citing a 1978 *American Journal of Obstetrics and Gynecology* (*AJOG*) editorial that questions the use of "gynecology teaching associate[s]" (GTAs), that is, women employed by medical and nursing schools to provide students with the opportunity to learn and practice pelvic exams with real, live bodies (Jamison 2014). The editorial relies on an efficient combination of discursive effects to contaminate and sexualize the role of the GTA. With a commiserative wink to the journal's then predominantly male readers, and with no small amount of the disgust that Kapsalis and I know well, the author exhorts: "My first question, as I suspect yours may be, was 'What *kind* of woman lets four or five novice medical students examine her?'" (8). Given that GTAs typically labor within a politics of feminist health care, and many (like Kapsalis herself) understand their role in the pelvic exam to be both instructional and performative, we can begin to apprehend the multivalent nature of the transgressions at work. What kind of woman indeed.

There are two relevant facts surrounding the bad taste in this physician's mouth: first, the majority of the novices in question in 1978 would have been male (Cooper 2003), and second, without professional models, these students would have examined either cadavers, synthetic prostheses, anesthetized patients, or "nonprofessional" women (typically sex workers) willing to be paid for such work (Ostrow 1980).[3] In short, live bodies had always been one option for medical schools but had previously not been authorized to speak (back) in the ways that GTAs are.[4] Kapsalis, a performance studies scholar, focuses her argument here on the "gendered spectator-spectacle relationship" in traditional gynecology, composed of "an active male physician-spectator and a passive female patient-spectacle" (1997, 23). In this configuration, there is no room for an *active female*, one that cannot be made absent, silent, or disabled by the masculinist prerogatives of institutional medicine (Ehrenreich and English 1978; Tsouroufil 2012). In her analysis, we can read the *AJOG* editor's disdain as a conservative reaction to the invasion of gynecological medicine by larger numbers of active female bodies. Indeed, by 1978 these bodies were not limited to the GTAs or consciousness-raised patients nurtured by the feminist health movement (Murphy 2012) but were also now female bodies intent on securing positions on the "physician-spectator" end of the spectrum.

But however obvious this editor's medical misogyny may appear, I want to deepen our interrogation of contaminated gynecology and sexualized medicine. Situated alongside the reactionary repression of *AJOG*'s editorial and Kapsalis's detached feminist irony, another voice reminds us that female agency and genital disgust are not mutually exclusive—that there are plenty of sexually active women who are reluctant to confront their genitalia. Attuned to this voice for many years, I want to add another layer to Kapsalis's analysis. In approaching the "what *kind* of woman" dilemma, we must move beyond the voice of institutionalized medicine to the disgust and repression that come from the patient on the table, to the *active female* who uses her recently acquired voice to cast suspicion on the nature of her own sexual body.

Speculum Practice

I did not learn to do pelvic exams with a GTA or professional model. In the graduate nursing school that I attended, students worked with each other to develop this skill and, I admit, it is a tricky business. Twelve of us were enrolled in a full-time program concentrated into one calendar year, and we came to know each other in the strange-yet-intimate ways typical of such conditions. And though long hours in lecture halls, clinical placements, and the library offered reasonable glimpses into our respective idiosyncrasies and bodily habits, they did not necessarily translate into using one another's vaginas for speculum practice. We also varied in the degree to which we wanted to perfect our techniques. Most of my classmates were not planning careers that would include pelvic exams; I, on the other hand, had been doing them for several years at the reproductive clinic where I worked before graduate school. Under these circumstances, my classmates and I negotiated our way through a new skill—and each other's bodies—with a mix of unevenly weighted performance goals, friendly familiarity, and the professional distance we were expected to maintain. And, as I recall, we did so without any undue sense of trespass.

I eventually became very good at pelvic exams, and I say this with the certainty that comes from uncomfortably witnessing those performed by less-skilled colleagues. Putting my patients at ease, both emotionally and physically, was something I took utterly seriously. I knew that I could almost always avoid provoking pain by taking the few extra minutes that slowing down and involving the patient required.[5] I was fully aware that these interventions would neither displace nor undo the discourses that surround "the GYN exam"—those of pain, martyrdom, and embarrassment that have been documented among women in the United States and elsewhere (Angier 2000; Domar 1986; Kaysen 2001). But I also knew that I could make those ten minutes into a distinct experience, one that could at best rival and at worst coexist alongside these more habituated experiences. I was rewarded for my efforts, not only by grateful (and sometimes astonished) patients who learned that it didn't "*have to* hurt" or be "awful" but also by a growing awareness that my *personal* relationship to this profoundly cultural process (Henslin and Biggs 1971) was in the midst of a transformation. Speculums became extensions of my body, and the expertise with which my hands became infused reflected an emerging body schema (Schilder 1950) that related to female genitalia with a new sensibility.

These embodied techniques reflect what Grosz (1994) and others (Berlucchi and Agliotti 1997; Carruthers 2008; De Preester and Tsakiris 2009) describe as a continually renewed body image. In her analysis of Lacan's imaginary anatomies, that is, those that emerge during the mirror stage of psychic development, Grosz asserts that "the stability of the unified body image . . . is always precarious. It cannot be simply taken for granted as an accomplished fact" (1994, 43–44). Some neuropsychologists suggest that "offline," or durable and relatively fixed, body images are primarily established by prior experiences that can then shape, though not determine, the acquisition of an "online" and more dynamic "representation of the body-part or sensation" (De Preester and Tsakiris 2009, 315). And though I was unaware of Grosz, Lacan, or the field of neuropsychology during the years I practiced, I nevertheless began to surmise that it was possible for bodily habit(u)s to change, and that this could occur with the conscious and sustained accumulation of experiences that resisted corporeal norms.

If, as I came to suspect, pernicious discourses of lack and disparagement could erode genital integrity, then perhaps a purposeful accretion of alternatively informed dialogues could fill these indurations, in the way that a deep and open wound heals through granulation. Indeed, Grosz contends that the "body image . . . must be continually renewed . . . through the subject's . . . ability to conceive of itself as a subject and . . . to be able to undertake willful action" (1994, 44). For De Preester and Tsakiris (2009), it is the "reservoir of accumulated representations of bodily experiences to which we have been repeatedly exposed" (315) that undergirds our "offline" bodily schemas. In allowing for the ongoing possibility that their gynecological encounter could be different, I offered my patients (and myself) a chance to renew, and possibly rehabilitate, the contaminated bodies in which we might have previously—and unthinkingly—resided.

But my skin sometimes felt tighter before it relaxed to accommodate my shifting sensibilities. For example, in many of the clinics where I worked, staff members could access medical services from the providers at either no or a reduced cost. This meant that not only could I become the patient of my clinician-colleagues, but that friends and coworkers could do the same with me. This was no longer peers-as-pelvic-models, however; I was now responsible for managing the care of these individuals, which often meant learning the details of their sexual practices and genital concerns. When I walked into my exam room one afternoon, fresh out of graduate school and in a new job, and saw the clinic receptionist—whom I did not know very well, whose desk I walked by countless times a day, who exercised a fair amount of power over my schedule, and with whom, frankly, I was still attempting to personally connect—I felt angry. It didn't seem fair to ask this of me, a staff member and a *brand new* clinician. I sensed that these visits were going to require yet another level of negotiation among bodies (genitals specifically) and social decorum, one that I wasn't necessarily eager to embody.

The happy ending is that I eventually developed a comfortable sense of mastery in these situations: my body, politics, and genital sensibilities coalesced to find more stable ground than the unease still provoked by conventional discourses about genital intimacy. My practice, increasingly informed by an ethic that was empathetic, egalitarian, and feminist, now helped me to imagine that the purposeful nature of my exams couldn't help but leave a distinctly informed residue in the body of the friend or woman on the table. But hovering around these behavioral acts, and circulating amid my practical efforts, was a question that feminist clinicians are singularly poised to consider: What *kinds* of friends or colleagues choose this particular brand of intimacy?

Writing this many years later, I can only speculate about the impacts that these encounters had on the other people who were involved; it is likely that free health care mattered far more than any inchoate genital integrity we might have been mutually tending. I can say, however, that in my case these bodily intimacies helped me to strike a more effective balance between friend and clinician, no matter how well I knew the woman before me. My verbal confidence came to match my clinical expertise, and I used every minute with my patients (often running over time) to provide them with information and critical questions. Contraceptives and sexually transmitted infections (STIs) were not just risk and benefit packages to be weighed, they were material gateways to the more abstract issues of sexual well-being and bodily integrity that fueled my desire to labor in gynecological health care. I got along well with my patients: they typically left my exam room with appreciably less tension and reserve, and it was unusual for my confidence to be shaken. But one evening, when an otherwise unremarkable[6] young woman looked right at me and asked, "God, how can you *do* this job?" I was momentarily helpless to answer her.

At the time of her objection, this patient was on the table in front of me and beginning to lie backward for her exam. We had conducted her health interview and not detected any major problems. We were going through the routine: I had stepped out and allowed her to undress in private, and she had donned her crepe paper gown and drape sheet. After listening at the door for the rustle of paper that indicated she had climbed onto the table, I quietly knocked, asking the rhetorical "Ready?" as I was already walking back into the room. I helped to arrange her feet in the stirrups, perhaps calling attention to the soft pads with which we made sure to cover them. I was always careful—vigilant even—not to introduce the stirrups until it was absolutely necessary, until I was ready to do the parts of the exam that dictate their use.[7] My fingers, therefore, were ready to work: gloved and lubricated, they were poised at the opening to her vagina. I was millimeters from her skin and prepared to perform a task with which I had come to associate more than a little self-respect.

"God, how can you *do* this job?"

I had not been expecting this.

Was I angry? Empathetic? Ashamed? Yes, yes, and yes—I was all of these things, most alarmingly the last. And I was silenced, if only temporarily. How to answer? What kind of woman *can* do this job? What *kind* of woman, other than a sex worker, earns her living by placing her fingers inside the genital bodies of others, an act that author and feminist gynecologist Christiane Northrup reports that her own patients find "disgusting" (1994, 241). After all, argues Katharine Young, "apart from physicians and lovers, access to the anal-genital region is specific to morticians and prostitutes, which suggests something about the body taboos that attend such access" (1997, 178). Encounters with female genitalia are so structured by proscription and contamination that all but the most heteronormative and clinically paternal are deemed suspect. Genital "ugliness," argues performance artist Joanna Frueh, "looms large in both cultural and women's consciousness . . . and a woman is likely to have a more medicalized than aesthetic consciousness about her own." (2003, 145).

As I relate this story now, more than twenty years after it happened, I am struck by an eerie recursivity, by how my patient's question, uttered somewhere around 1992, both anticipates and succeeds the two variations posed earlier, those of the *AJOG* editor and Kapsalis; questions that want to know, more generally, "What is *wrong* with these people?" In the cumulative model of female genitality posited in this chapter, these three questions are chronologically—and discursively—stratified across overlapping cultural zones. Kapsalis's feminist challenge (What kind of woman studies pelvic exams?) may resonate better with contemporary readers than the threatened patriarchy of *AJOG* (What kind of woman gives multiple medical students access to her body?), but they both index the disquiet provoked by genital encounters. My patient's provocative query neatly interleaves itself between the other two and resonates with an all-too-available reality. We may not feel the need to stop our gynecologist in her tracks with an utterance of disgust, but we do understand why Kapsalis feels the need to acknowledge the transgressive nature of her project. These three interrogative events index complex and concomitant layers of genital discourse, the figurative ground upon which the lived experience of vulvar pain is unhappily situated.

Kapsalis reminds her readers that "cultural attitudes about women and their bodies are not checked at the hospital door" (1997, 63). This book departs from that insight to argue that vulvar pain cannot be apprehended adequately without exposing and interrogating the cultural attitudes that surround and construct female sexuality. In making this claim, I locate myself among critical scholars of embodiment who suggest that "the specific cultural meaning of . . . bodies is not distinct from but deeply embedded in the relations of domination . . . that have framed" those bodies (French 1994, 72). The habitual and hegemonic disavowal of the vulva, the "absence [that] . . . is as psychically invested as its presence" (Grosz 1994, 41), efficiently functions as a depositor of cultural residue in the female sexual subjectivities constituted by its accompanying discourses. Povinelli (2006) argues that discourses can "make and unmake . . . bodies" through a "politics of cultural recognition" (22–23). In this book, I investigate female bodies (un)made via vulvar disavowal and (mis)recognition and argue that discursive and reiterative erasures, rather than producing a clean corporeal slate, accumulate as shameful sediment, an intractable film held in place by a deep cultural ambivalence toward the proper "place" of the nonreproductive female sexual body.

STORIES

Frances

Medical assistants were usually the first to see completed medical histories; their practice was to skim the forms while they walked with patients down the hall and gathered vital signs in the exam rooms. One morning, as Gia handed off a chart to Dr. Robichaud, she gave the physician the kind of look that meant, "You might want to read this one a little more carefully before you go in there." Upon reviewing it a few seconds later, we saw that the patient had written the word "disfigured" into the symptom section. Dr. Robichaud said simply, "This is bad."

Gia also told us that this patient had brought along a couple of diaries that documented her symptoms in great detail. These diaries—painstakingly kept and filled with both the vastness and minutiae of vulvar pain—were a physical index of the existing gap between a woman's lived experience of her disease and the much narrower lens through which it was viewed by her health care providers. No matter how many pages had been filled (in this case, it was two spiral notebooks), the clinician would usually base most of her treatment plan on little more than ten minutes of interview details and—most importantly—the findings of a physical exam. Patients who coped with invisible symptoms by inscribing them onto the written page often found that their provider's interest in their pain stopped at the physical receipt of these sheets of paper. Having scrupulously amassed the evidence of their disease, patients found that they could not subsequently give it away, could not transfer its symbolic reality into the hands of their caretaker. Rather, these documents were typically slated to remain in the permanent collection of their genital pain.

The patient in question—Frances Hoffman—had accumulated more than diary pages in the three years that she had been symptomatic. Like Louise, she found it almost impossible to sit for long periods of time, and so she had acquired an inflatable "ring pillow" that took the pressure off her vulva. Frances told me later that she had purchased an attractive bag to make her pillow more portable. Due to the type of "down there" disability that her pillow conveyed, however, she was embarrassed to use it in certain situations (such as her first week at a new job). On this particular day, the clinic fell into that category, and so when we met Frances, she was standing up to avoid the uncushioned contact of her vulva with the built-in bench in the exam room.

Unique among the women I met at the VHC, Frances's pain was primarily the result of a surgical complication—specifically, a nerve tumor (*neuroma*) that developed after she had a vulvar cyst removed. Since neuromas can develop in any area rich in nerve fibers, there was nothing particularly vulvar about her symptoms, other than their anatomical location. But part of why Frances was in unremitting pain three years after her surgical complication was because she had not located a physician invested enough in her vulva to take her complaints seriously. As I observed what turned out to be a lengthy and emotional consultation, I came to appreciate not only the investment that Dr. Robichaud had in Frances's vulva but also her understanding of the imbricated and pernicious layers in which Frances's pain was lodged.

In addition to a ring cushion, a new bag, antidepressants, pain medication and antibiotics (all ineffective), sleeplessness, lidocaine patches, and several volumes of recorded data, Frances had collected an enormous amount of shame about her symptoms. This was because she felt as if she had been "railroaded" by a "cocky" surgeon into having a procedure she may not have needed. Her vulvar cyst—called a Bartholin's cyst—is a benign condition, problematic only if it becomes infected. This had happened several times in Frances's past, but she had always been treated with antibiotics or a simple procedure that drained off the infected fluid. After a woman turns forty, however, an infected Bartholin's cyst carries a slightly increased risk of malignancy, and cautious gynecologists recommend excising them rather than simply treating the inflammation (Droegenmueller 1992; Omole, Simmons, and Hacker 2003).[8] When Frances's cyst recurred in 2002, she was not only over forty, but she had just relocated from one state to another; her indecision about a surgery that she had previously refused (some women opt for excision with the first infection) was now further confused by the added fear of cancer and by the fact that she had established few relationships—medical or otherwise—in her new city.

Sufficiently unnerved, Frances complied with the recommendations of the "cancer specialist" with whom she had fortuitously secured an appointment (an acquaintance had recommended him and she got in on a cancellation). "He gave me the bum's rush. I didn't trust him. I shouldn't have let him rush me into it. [But] I didn't know anyone to ask. He said 'cancer,' and I was scared." Dr. Robichaud told me later that the procedure to remove these cysts was "notoriously complicated," sometimes referred to as "the bloodiest little surger[y] in gynecology." Because of the proximity of the Bartholin's gland (the source of the cyst) to several bundles of pelvic nerves, it was tricky to avoid complications like the one that Frances had sustained (Leclair and Jensen 2005); the anatomical distortion produced by an inflammation would only compound the risk. Because of this, Dr. Robichaud always tried to treat the infection before she performed surgery. Telling Frances "I usually try to cool them down first," Dr. Robichaud diplomatically communicated that this might have played a role in the development of her neuroma.

In drawing this contrast, and in frankly asserting "You have a bad complication" following her physical exam, Dr. Robichaud attended to Frances's vulva in the manner of a vulvar specialist; that is, her interest in Frances's vulva was not circumscribed by its risk for acquiring a malignancy, infection, or other pathology. Rather, Dr. Robichaud promoted and preserved vulvas that were "robust," "juicy," "supple," and alive—vulvas that were vibrant in and for themselves and with which women could develop a wide range of relationships. Her choice to let vulvar infections "cool down" before she excised any part of them reflected her investment in the well-being of her patients' genitalia, as did her habit of allowing women over forty to trade a risky surgery for closer clinical follow-up. With these decisions, Dr. Robichaud expanded the definition of genital integrity beyond the mere absence of cancer or disease to one marked by a maximum of anatomy and pleasure and a minimum of pain and shame.[9]

Dr. Robichaud determined that Frances had some concomitant *vulvodynia*, and the two women decided on a treatment plan of PT for the pelvic floor tenderness that had developed thus far. She was clear with Frances that there was little she could do for the neuroma pain, however, aside from systemic neuromodulators (taken as pills), all of which had the "typical side effects" of fatigue, wakefulness, "cloudy" feelings, headaches, and gastrointestinal changes; "nothing," in Dr. Robichaud's words, "that makes anyone say, 'Sign me up!'" Frances declined these drugs for the time being after describing how sensitive she was to medication, but she was satisfied and even hopeful regarding her overall treatment plan, particularly about the difference that PT might make in her eventual pain level. She made plans to follow up with Dr. Robichaud three to four months later, and we all began readying to leave the room and say goodbye. What happened in the next few moments, however, brought our movements to a grinding halt; it was also an unparalleled ethnographic moment, the kind I had expected to witness in the field. Though I knew that patients could have profoundly different experiences with providers who were attuned to their symptoms, I could never predict exactly how this would occur, and Drs. Robichaud and Erlich routinely surprised me with their ability to locate traumatically infused layers of experience that were not always readily apparent. In this case, it was the dense network of sexual and bodily shame, rather than nerve fibers, in which Frances Hoffman's neuroma was firmly embedded.

Dr. Robichaud referred Frances to one of the clinic's most trusted physical therapy affiliates, a group of women named Cathy, Hanna, and Joy. Aware that there was likely even more to Frances's story, she stressed that these therapists provided an "extra piece" of emotional support to their work and then asked Frances if she "blame[d] herself" for what had happened to her vulva. As if on cue, Frances began crying ("hard," according to my notes) and said "YES! And that's the hardest part of this!" As she continued to share her sadness and regret with us, Frances began to unburden herself of the shame and self-blame that she had been carrying for three long years.

Through her tears, Frances repeatedly stressed that her surgeon "didn't listen!" when she began calling him with concerns immediately after the surgery. (He apparently put her off for close to ten days.) In considering the nature and consequences of that particular erasure, I knew that listening supportively *now* was the minimum that I could do for this patient. But as

I made eye contact with Dr. Robichaud, and ascertained that I had a role to play in this unfolding drama, I saw that she was taking things a step further by sitting back down on the exam room stool. At that point, we had been with this patient for over one and a half hours; Dr. Robichaud knew that several patients were "roomed" and waiting for her, and she had already elaborated a complete and supportive plan of care for this one. And yet, as Frances described the scope and the impact of her surgeon's disavowal, Dr. Robichaud, rather than taking a gracious (and understandable) leave, chose instead to add another layer to Frances's story: her concerted and *bodily* attention to Frances's grief. Rolling with the stool toward Frances, she gently told her: "Sometimes you don't get over something, but you learn to live with it. You're grieving. You're sad and you're mad. And you have this daily reminder."

Dr. Robichaud is a sensitive clinician, and her clinical style would likely be marked by exchanges of this emotional caliber regardless of the specialty she chose. But her behavior in this story, her ability to recognize and facilitate Frances's bodily grief, signifies a cultivated attunement to the needs of vulvar patients. Like the Vietnam grunts in Tim O'Brien's poignant short story "The Things They Carried" (1990), symptomatic women carry mountains of "intangibles" along with their pill bottles, ring cushions, and lidocaine gel. O'Brien's grunts are loaded down with rifles, pocket knives, Kool-Aid, cigarettes, steel helmets, and sewing kits; they also

> carried all the emotional baggage of men who might die. Grief, terror, love, longing—these were intangibles, but the intangibles had their own mass and specific gravity, they had tangible weight. They carried shameful memories. They carried the common secret of cowardice barely restrained, the instinct to run or freeze or hide, and in many respects this was the heaviest burden of all, for it could never be put down, it required perfect balance and perfect posture. (15)

In the clinic, these intangibles existed as a dense and palpable fog, a thick miasma of alienation and despair that could be either circumnavigated or carefully waded through. Dr. Robichaud—as exemplified by this moment with Frances—was committed to guiding her patients both through and away from feelings that she knew could too easily lead to stagnation. By being open to, and by initiating a dialogue about, her patient's grief, Dr. Robichaud is mindful of Frances's intangible burden and offers her a place to put it down—at least temporarily.

An investment in vulvar well-being and an attention to its vulnerability are mutually constitutive; they are also predispositions that can be actively cultivated (Carruthers 2008). Dr. Robichaud's ability to locate self-blame, despite an (apparently obvious) outside perpetrator, has been honed by her work with vulvar pain because of the other genital and sexual stories with which it articulates, the genital baggage under consideration here. As it happened, Frances's story contained explicit traces of the insidious trauma that Root (1992) has argued is an underlying current of most women's bodily habitus. Frances told us that she had two important people in her life at the time of her surgery—a boyfriend (her husband at the time of her clinic visit) and a male best friend, who was part of the reason she had recently moved. Each was available to help her that day—her friend in the morning and her boyfriend in the afternoon. Although he couldn't stay, her boyfriend wanted to be there when she checked in to the hospital, and so both men were (temporarily) present when the surgeon came in that morning to review the procedure.

During this part of our conversation, Frances reiterated how arrogant her surgeon was, even perching one leg up on the exam room stool and splaying her hips in order to demonstrate the "cocky" way that he occupied the one in his surgical suite. What upset her the most, though, was the question that he posed to her as she surrendered to the anesthesia. While dreamily counting backward from one hundred, she heard her surgeon ask, "So Frances, which one is the boyfriend and which one is the best friend?" In retelling this episode three years later, Frances could imaginatively counter her surgeon and say to us, "So I brought two men, so what?" But what lingered, mixed with the self-blame she still carried about what would more than likely be a lifetime of genital pain, was that she recalled feeling "ashamed and dirty" about her surgeon's comments. Losing consciousness on the operating table, and still nagged by misgivings over both the procedure and the surgeon himself, Frances recounted that the "last thing [she] was aware of going into this surgery was this shame."

When Dr. Robichaud asked Frances to describe her pain during the physical exam, she initially answered with descriptors that I heard frequently in the vulvar clinic—it woke her up, she didn't sleep for a year, it felt bruised, and it hurt to touch it. She then continued in slightly more reflective terms: "The area feels . . . it doesn't feel like my vagina. It feels like it's trying to hide. The labia minora—I used to feel it hanging, I could feel my clitoris. It was ticklish to touch. Now, it feels like it's pulled in, like it's hiding." That Frances's genitals have gone underground makes profound sense. Her surgeon's comments, the bulky ring cushion, her inconvenient complication, my (initial) impatience with the drama of her story, and even Gia's knowing look, are the stuff of which Frances's shame is made, the cultural habit(u)s through which we disavow her vulva, through which we try to make her sexual body disappear. Frances's labia and clitoris are "pulled in [and] . . . hiding" in order to protect themselves from these acts of disparagement, like the muscles of the pelvic floor in their efforts to avoid penetrating interlopers. What the vulvar clinic did for Frances was to bring this sexuality into the light of day. Dr. Robichaud made efforts to listen, to *sit down* in the face of a story of erasure, to lighten the load of this repression and hiding, and provide an opportunity for something that was rapidly and permanently closing down to open up and even flourish. As was sometimes the case, Dr. Robichaud had little in the way of pain relief to offer this patient. But that day she gave Frances her vulva (back), along with the promise that follow-up visits would continue to coax Frances's genitalia out of hiding.

There is no doubt that Frances's body, along with those of her symptomatic cohort, had accumulated an excess of unwanted genital experiences. And because they were extraneous to her vulvar pain, these are experiences with which many women can identify. But Duden (1998) reminds us that bodily imaginaries, while general in some respects, are also contextually specific, and that the objective continuity of physical bodies is disrupted by both time and space. And though I do not agree that "giving a history to the female body" (vi) necessitates a complete and mutual break, in other words, between "their" bodies and "ours," I am convinced that the contextually specific bodies that Duden insists we recognize help us to better understand what Drs. Robichaud and Erlich create for their patients through the alternative imaginaries that the clinic allows them to shape.

When Dr. Robichaud says to a patient named Joan, for example (of her *lichen sclerosus*), "You don't have a vaginal problem, you have a vulvar problem. That's like saying an arm instead of a leg," she provides this patient with another layer in her genital reality; she uses their time online to adjust Joan's offline genital imaginary (Carruthers 2008). For a woman like Joan, who had lost almost all of her clitoral and labial contour, these purposeful and clear sentences infuse a disparaged and ignored bodily hexis with neutrality and concerted attention. They create and deposit the idea that it's both acceptable and even worthwhile to, in Joan's words, "go down there and look." Both clinic physicians with whom I worked understood and consistently disrupted the social forces they were up against by promoting greater genital awareness through medical instruction and intervention (e.g., "See this diagram? This is your vulva."; see figure 2). Minus these interventions, collective genital proscription and defamation layer themselves under, around, and on top of whatever personal resources a woman may have otherwise acquired.

Importantly, the genital ease of a provider and the *dis-ease* of a patient are not mutually exclusive; rather, they are most constructively understood in the imbricated and sedimentary terms theorized here. Much like the Bartholin's fossa (the source of Frances's cyst)—a functional and usually healthy drainage duct nestled deep within a network of nerve fibers—the individual and therapeutic acts of Drs. Erlich and Robichaud can wedge themselves into the deep-tissue discourses of inconsequence and create new bodies for their patients, bodies that open up on the exam table and exclaim "I can see!" as they are handed a mirror and examine their vulvas for the very first time. Divested of their burdens—even momentarily—the complex striations of symptomatic women's dis-ease become more visible, reconfiguring and making them more "manageable" for clinicians and (feminist) theorists alike.

Deirdre, Lily, Isabelle, and JoJo

> DR. ROBICHAUD: So, tell me what's been going on.
> DEIRDRE: It hurts.
> DR. ROBICHAUD: Where?
> DEIRDRE: Down there.
> —From my field notes

Deirdre came to the clinic with her mother, Jan. She was seventeen years old and had collected a number of difficult experiences in her young life. This included being sexually molested by her mother's ex-husband between the ages of two and six and a belief that her vulvar symptoms began around that time. Deirdre had also acquired a counselor along the way, a woman who told her that "it might be some PTSD stuff in [her] head." Prior to her visit with Dr. Robichaud, Deirdre had been treated for several STIs, including one that came to compromise her future fertility. Jan asked if the clinic could do a pregnancy test that morning because she knew that Deirdre had been having unprotected sex with her boyfriend. (My field notes say: "I am reeling as we take all of this in, trying to find places for the vulvar pain to fit. Dr. Robichaud is not batting an eye.") After gathering the rest of Deirdre's history, performing a physical exam, and determining that Deirdre's pain was "classified into four different pains," Dr. Robichaud began outlining the details of a treatment plan. She soberly punctuated her clinical monologue by saying, "I think you have some hard work ahead of you."

Long before this extra mantle was added, I had begun to feel my own body slouching with the weight of this young woman's burdens. Later that afternoon, I wrote: "This mom

[is] . . . probably in her early forties, but looks older, like she is tired and has had a hard life. I wonder, as she relates [Deirdre's] history so matter-of-factly, how a mother and daughter ever recover from something like this. It feels much too big for this small exam room, for us four bodies to hold." These burdens, what Bordo (1993) refers to as the "unbearable weight" of gendered asymmetry, both index and sustain the fragile and disjointed relationships that many women have with their genital bodies. The fact that their vulvas are an inappropriate topic for virtually all conversations, save the one their physician now expects them to have, undergirds much of what these patients accumulate along the way to a diagnosis and treatment of their symptoms.

Many of the women I met related stories about the "well-intentioned [and] often contradictory" (Kempner 2014, x) advice and treatment they received from inexperienced providers. The frustration and disappointment associated with these false starts often led them to drag their feet about making an appointment at the specialty clinic, even after they became aware of its existence. Lily, for example, told me about a clinician who literally threw up her hands and said, "I don't know what's wrong with you!" right before she referred her to Dr. Robichaud. It was also not uncommon for patients to be cajoled into finding ways to "ride out" the pain: Isabelle was encouraged to simply "smoke a little grass." Indeed, advice to "stick it out" was rampant among the gynecologists that my informants had consulted. Medical recommendations to "have a glass of wine" or "wait until you have a baby" were generously distributed by providers whose understandings of female sexual pleasure included an acceptable amount of pain.

The professional vulnerability that nonexpert providers refused to acknowledge, regarding their inability to provide a diagnosis or cure, precipitated significant psychic and material burdens for patients who felt compelled to pay for and follow misguided treatment plans. Lily, for example, had collected "an abundance of medication" from the doctor who ultimately threw up her hands in frustration, while JoJo, whose physician held fast to his belief that drinking wine would "loosen her up," purchased more bottles than she could recall, hoping to find a variety she enjoyed. The fact that she did not "even like wine" mattered little in the context of complying with advice that she both sought and received with sincerity. Moreover, the relationship between inadequate clinical expertise and pernicious assumptions about female sexuality—how a *relaxed* pelvic floor articulates with ideas about *loose* women—merits our scrupulous attention, particularly when it is voiced by patients and their partners. Upon hearing Dr. Erlich explain how PT would soften his wife's pelvic floor muscles, one husband responded by asking, "So, does that mean I have to go out and get her drunk?"

The discourses through which these (clinical) behaviors can be at least partially understood are social realities that weigh heavily on patients with genital pain. The singular experience of constituting a clinical "puzzle" for one or more inadequately educated doctors meshes far too seamlessly with the collective experience of living in a body that has always mattered too little. A woman's genital familiarity—in any amount—is easily threatened when the experts charged with her care demonstrate more clumsiness than proficiency. When your doctor proclaims, after a cursory examination, "Well, you've got lube, so you're okay," the matter of your sexual pain and pleasure is positioned firmly behind the matter of your sexual function.

These initial encounters with medicine are based upon, permeated by, and ultimately productive of discourses through which the female sexual body is deferred and rendered absent through its inscrutability. A chronic condition, this absence resonates deeply and palpably, in the way that a phantom limb is infused with a ghostly corporeal presence. In the contemporary United States, women with vulvar pain index the discursive amputations of their sexual bodies, and they ache with the burdens of inconsequence and erasure. They perform important cultural work in their efforts to make their experience medically known, but the weight of these cumulative encounters slows their individual progress toward adequate treatment. For some, slowness turns to stagnation, and by the time they reach the clinic, they are almost too heavy to walk through the door.

DEEP-TISSUE DISCOURSES

Anything Wrong before I Touch?

Few patients referred to the VHC by nonspecialist providers had ever been sent to physical therapy for their vulvar symptoms; if they had, it had rarely diminished their pain in an appreciable way. But PT was an early staple in the treatment plans of Drs. Erlich and Robichaud, given the number of women who were reluctant to have surgery right away. Added to the fact that most patients' insurance plans covered the cost of PT (and not sex therapy), it made both practical and clinical sense for the clinicians to invest substantial effort into linking patients with someone who understood the particular needs of vulvar pain.

Physical therapy is intended to help a woman better understand and control the musculature of her pelvic floor (Gentilcore-Saulnier et al. 2010; Hartmann 2010). Vulvar pain, almost always superficial, becomes deeper and more confusing as the surrounding musculature develops a compensatory response. With the approach of a potentially painful stimulant (e.g., tampon, penis, or gynecological instrument),[10] women with vestibular pain in particular learn to pull in their vaginal and pelvic floor muscles in a protective "tail-tucking" maneuver. Although this behavior may be technically volitional the first few times it occurs, by the time a patient has developed *pelvic floor myalgia*, it has become a bodily hexis over which she has more or less lost control. Ultimately—and unsurprisingly—this strategy proves maladaptive, as the once superficial, acute, and localized vestibular pain becomes deep, diffuse, and lingering in the large muscles of the pelvic floor. PT, done by a trained and empathetic provider, teaches a woman with pelvic floor myalgia to undo, or at least gain greater control over, this compensatory pain (see figure 6).

This deep-tissue pain is also known as *vaginismus*, a Freudian term used to describe vaginal clenching believed to be related to unresolved psychic trauma, usually regarding heterosexual coitus (Kessler 1988; Payne et al. 2005). Some physicians, including Dr. Erlich, object to the neurotic and

FIGURE 6. Muscles of the pelvic floor

The Muscles of the Pelvic Floor

Puborectalis
Pubococcygeus — Levator ani m.
Iliococcygeus

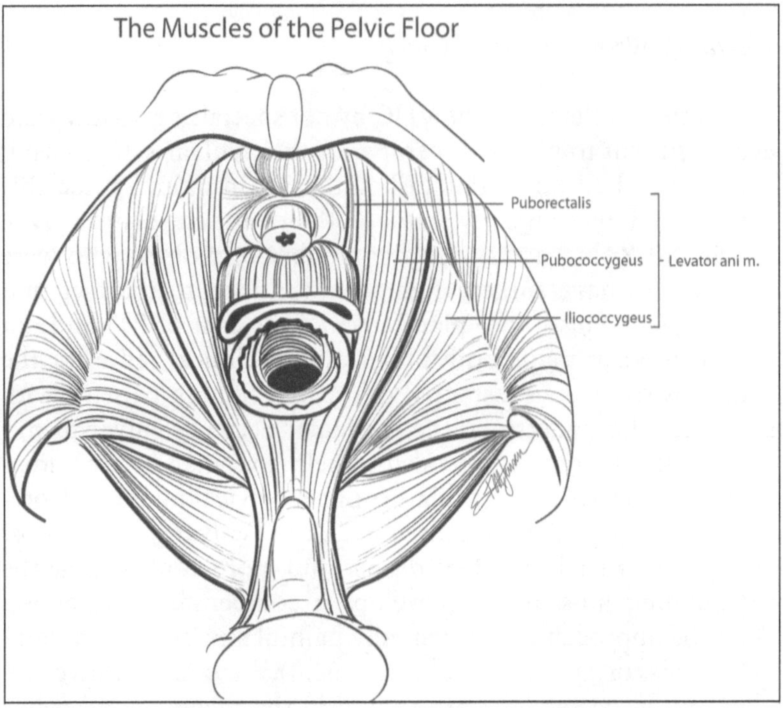

Image used with permission of Robin Jensen.

psychosomatic connotations of this term (Leclair and Jensen 2005; Reissing, Binik, and Khalife 1999) and describe pelvic floor pain with the more clinically neutral term "myalgia" (literally, muscle pain). With less symbolic baggage, the term myalgia allows an alternative set of correlational or causal factors to be considered as part of a patient's treatment plan. Communication among physicians, hospitals, laboratories, and insurance companies is facilitated through a glossary of nationally recognized billing codes that correspond to diagnostic categories. In short, medical diagnoses (e.g., upper respiratory tract infection) are assigned numerical codes, which are then used to determine a range of associated practices: number of visits needed, appropriate treatments, expected cost of recovery, and so on. Part of the early work of the National Vulvodynia Association (NVA), the patient advocacy group affiliated with female genital pain, was to get these codes established in order to medically legitimate symptomatic women. During my fieldwork (2004–2005), the research hospital in which the vulva clinic was housed was using two separate billing codes for the condition that made PT services reimbursable, one for vaginismus and one for pelvic floor myalgia. These multiple codes demonstrated not only the shifting nature of vulvar pain diagnoses (NIH 2012) but also the feminist politics at play within the clinic. Dr. Erlich, for example, told me that her choice to check the myalgia box on the patients' billing forms was as ideological as it was clinical.

Physical therapists who specialize in treating myalgia often have a background in another variety of pelvic floor work: the muscular retraining of women afflicted with urinary incontinence. Indeed, PT techniques have proven so effective in managing incontinence that most gynecological urology practices now include at least one of these specialists (Berghmans et al. 2000; Pages et al. 2001). PT became a part of the vulva clinic's toolkit because Dr. Jenkins, the clinic's founder and former director, approached Cathy, whom he had heard was doing incontinence work, and asked if she and her partners would be willing to learn about vulvar pain. ("He literally found us in the phone book," Cathy told me one afternoon). His reasoning was a simple case of reverse mechanics. Women with urinary incontinence are taught to strengthen and direct the muscles of their pelvic floor, in other words to contract and release them according to their physical needs (containing or allowing a flow of urine). It made sense to Dr. Jenkins that women whose muscles maladaptively *contracted*, rather than loosened, could be taught to exert that same kind of corporeal control. Physical therapists and gynecologists worked together to establish a few initial protocols and criteria, and willing patients and their therapists began to experiment with techniques.

Many women have built up so much muscular tension during their undiagnosed years that they experience vaginal and pelvic floor pain in addition to penetrative resistance.[11] They often do not perceive this pain until a partner attempts (or achieves) penetration, but the amount and the quality can change dramatically over the course of months or years.

Women whose pain was only experienced at the point of vestibular (skin) contact, and who could tolerate coitus (or a pelvic exam) "if he could just get past that point," eventually ended up having deep and lingering pain, sometimes for days after vaginal penetration. An early diagnosis of VVS, and a timely prescription and routine use of topical lidocaine, can conceivably prevent the development of this deep-tissue involvement, if the vestibular skin can be managed well. Women describe their muscular pain as "burning," which contributes to significant amounts of bodily confusion, in that their vestibular (skin) pain is often of the same quality. It eventually becomes nearly impossible—even for the most informed women—to differentiate between the two (or more) varieties of pain, leading to an anatomical conflation and overall sense of disorder about "what all's going on down there." Diagnostic precision can be tedious and taxing to achieve, and Drs. Erlich and Robichaud spend a good portion of the physical exam trying to teach their patients how to distinguish the different sources of their pain.

There are several methods employed by these pelvic floor specialists: bioskeletal evaluation and alignment, strengthening and stretching the muscles that surround and support the pelvic floor (abdomen, hips, thighs), breathing and relaxation exercises, dilator (home)work, craniosacral therapy, vaginal and rectal myofascial (or trigger-point) release, and biofeedback (Bergeron et al. 2002; De Andres et al. 2015; Glazer and Rodke 2002; Reissing et al. 2005). These techniques are used in various configurations by individual therapists, and it is probably safe to say that no two PTs approach pelvic floor pain in exactly the same way. When I researched the providers on the clinic's referral list, for example, I spoke with several who did not view internal/vaginal work as a necessary component of their treatment plan and others who were not interested in what biofeedback had to offer.

In the most general terms, physical therapists are split into two camps: those who understand pelvic floor pain as a musculoskeletal or alignment issue, best approached with whole body maneuvers, regional strengthening, and complementary breathing exercises, and those who, by contrast, work more from the inside out, concentrating on the hands-on release of tightly contracted pelvic floor muscles surrounding the vagina (and sometimes rectum), complemented by supportive and whole-body techniques such as diaphragmatic breathing or craniosacral massage. This second approach is nicely illustrated by an exchange between Cathy and Libby:

> CATHY: [Moves to a new spot in Libby's vagina; a new place for pressure. She says she can feel some tightness.] As I hold, is that discomfort more, less, or staying the same?
> LIBBY: Pretty much the same.
> CATHY: [Describes what she's doing with her fingers.] There's turning, and now pressure on the right.
> LIBBY: OOH!
> CATHY: Just keep breathing—finding those places. How are you doing? Can [you] do one more? [She describes and then performs a horseshoe shape around the vagina, from one side to the other.] Good job, you're breathing. That's it. I can feel some of those tight areas. They're just leftovers from years of pain. They're flexible, they're soft. You just work them out until they get soft. This is where the work is. There is where the work is.
> LIBBY: Yes, I can feel it.
> CATHY: If you can't relax where the discomfort is, you relax around it, or as close to the discomfort as you can get.

More than a few of the women that I met through the clinic were averse to—even repulsed by—the idea of PT that would take place "down there." Clair, who was one of these women, was in her mid-fifties and in an actively Christian marriage with her husband Dwayne when I met her. At the time of my fieldwork, Dr. Erlich had diagnosed her with VVS, and she was considering surgery. Though extreme among my informants, Clair's reluctant narrative differs from that of other informants more in degree than kind and captures many of the reasons why symptomatic women hesitated to pursue PT despite their physician's strong recommendation. I reexamine this issue in chapter 5, paying greater attention to the geographic and socioeconomic factors that contributed to these modest compliance rates. But here, in my discussion of bodily accumulation, Clair provides us with an excellent feel not only for the content of PT sessions but also for some of the bodily striations with which these sessions awkwardly and poignantly interleave.

> CLAIR: I even tried PT, and that was a horrible experience.
> CHRISTINE: Tell me more about that—so one of the doctors referred you to a physical therapist?
> CLAIR: Mmm-hmm.
> CHRISTINE: And what did they tell you was wrong, what did they say the PT was for?
> CLAIR: The same, they really had no um, um, . . . I believe, I believe that she had a name for it.
> CHRISTINE: Did she say vaginismus?
> CLAIR: Yes, yes.
> CHRISTINE: Did she explain what that was to you?
> CLAIR: She said if I do these exercises, I'd feel better, but that didn't work either.
> CHRISTINE: Tell me about, what did she want you to do, and how long did you try . . . ?
> CLAIR: Well we tried sitting on this great big ball, we tried relaxing, and, um, a lot of exams, and she'd put a mirror down there and show me, and she, um, wanted me to, um, uh, . . . satisfy myself and . . . all of those things. And . . . the *whole* experience was uncomfortable to me. It, it was, it was um, . . . I don't know, it just, it just um, I guess it was my generation. We just are not *used* to having our legs spread in a mirror and two women talking about it and looking down there and all this *exposure*. And I was never comfortable with it. *But* I was willing to go through the humiliation of it if it worked. But it didn't! So I said, you know, this is . . .
> CHRISTINE: Yeah, and how many times did you go, do you remember?
> CLAIR: I think I went three times to her. And she'd just

act like I, . . . We were talking about my finger! And it wasn't like that to me. So . . . you know, it's, uh, maybe a generation problem I don't know, but . . .

CHRISTINE: You felt, um, vulnerable or exposed, or sort of . . . ?

CLAIR: I felt exposed, vulnerable. I just felt like . . . um . . . I don't know, it felt similar to, um, . . . sexual abuse . . . situations, that I'd had in my past. Um, . . . it just didn't feel, um . . . From a spiritual aspect, 'cause I'm a Christian, it didn't feel *right*.

CHRISTINE: Mmm, hmm, mmm-hmm. So people use the word *safe* when they talk about that. Like, did it feel unsafe, or, it more didn't feel right?

CLAIR: It just didn't feel like . . . It was, it was *too much*. It was just too much . . . intimacy with another woman. I, I . . . That's the only way I can explain it. But yet I can have a female doctor examine me and I don't have the same feeling.

As I hope I have begun to make clear, PT is not for the faint of heart. For virginal VVS patients in particular, establishing a connection between penetration (e.g., of the therapist's finger, or a therapeutic dilator) and pain *relief* was a daunting task. This challenge proved even greater for women like Clair, whose symptoms were—at a minimum—obliged to a history of sexual abuse (Cvetkovich 2003; Harlow and Stewart 2005), a set of religious convictions, and age-related ideological beliefs that she referred to as her "generation." Clair described to me how she would "just freeze" and pretend to be asleep when the man who molested her got into bed, the bed of a couple for whom she babysat, and in which she was invited to lie down after the kids fell asleep. Physical therapists in both camps (musculoskeletal alignment or pelvic floor) make the connection between the bodily "freezing" of that little girl and the tension in Clair's adult pelvic floor; those in the latter camp, however, make emotional connections that, while not necessarily running deeper, function instead like curious tentacles, searching for the idiosyncratic sources of a woman's "holding pattern." Given my own suppositions about Vulvar Disease, I felt a greater affinity toward this second group, as their therapeutic trajectories allowed for both a "muscular" version of psychic pain (Harlow and Stewart 2005; Wilson 2004) as well as a set of cultural causes of pelvic floor pain.

For women able to commit to PT, the results were almost always successful, although, as with many bodily afflictions, successful outcomes were often contextually defined. Through physically intimate and supportive sessions, patients were encouraged to directly confront their genital bodies and sexual selves and to connect those facets of their identities to the rest of their worlds. Hanna, who routinely instructed her patients about the more sociocultural aspects of the bodily hexis, told Daphne during one session that the biofeedback they were about to begin using translated into "how you do life." Because of this explicit connection, and because of an experience that I had with my own body, I want to conclude my discussion of vulvar pain's accumulated nature with a closer examination of this therapeutic technique. My personal biofeedback session epitomized my role as a participant-observer of this disease condition, by both strengthening the hypotheses I held about symptomatic women's bodies and disrupting some of the beliefs that I held about my own.

Psychologist Howard Glazer and physician Gae Rodke were two of the first clinicians to advocate using biofeedback as a treatment for vulvar pain. In their 2002 patient-centered text, *The Vulvodynia Survival Guide*, they define the technique as

> an electronically assisted measurement of physiological processes, such as heart rate, blood flow, and muscle contraction. Through the use of highly specialized computers, a specific physiological process is translated into an auditory or visual signal so that the patient can learn to control it—and return that physiological process to more normal, stable, healthy levels. (63)

In more general terms, biofeedback is information about the body (and its world) provided by the body. Because of the "fight or flight" responses of the sympathetic nervous system, for example, most of us experience notable physical reactions in dangerous situations, including an increased heart rate, palpitations, or altered breathing. These biological events alert us to the changed nature of our physical situation, offering us the opportunity to move or change our body's behavior—fight or flee. These symptoms become associated with danger so that if and when they reappear (e.g., in the context of a police siren or a public speaking engagement), we understand these new situations through the same interpretive framework: I must be endangered because my heart is pounding. This, in itself, is biofeedback—my body is giving me information about how I interpret my world.

Clinical biofeedback gives this set of concepts a therapeutic spin. In a social context where performing a wedding toast, for example, does not constitute the same danger as a grizzly bear on your hiking trail, it is helpful to develop bodily skills to alter inappropriate physiological responses; those corporeal behaviors (e.g., diaphragmatic breathing) then come to be associated with the new, relaxed bodily state. This is manipulated biofeedback. In the physical therapist's office, and as described by Glazer and Rodke (2002), this manipulation is carried out with the help of dedicated equipment and software that complement a patient's developing skill set (and bodily repertoire) with visual and auditory cues.

A good deal of the PT that I observed was done by the practice that consisted of Cathy, Hanna, and Joy. This was the group that Dr. Jenkins had first contacted, the one with which the physicians had the most experience, and the local group that made the widest connections between physical and social bodies. They used all the techniques that I previously listed, and they used biofeedback with all of the vulvar pain patients whose sessions I was invited to observe. I got to know these PTs first through Libby, whose treatment lasted almost six months (averaging two treatments per month), and then subsequently through both Daphne and Julia R.[12] Each of the three therapists had slightly different styles, but they utilized biofeedback in the same way: they attached electrodes to some part of the patient's body (jaw, inner thigh, and around the chest or diaphragm) and then connected them to a monitor. Patients could then read their bodies' rhythms via color-coded lines that were visible on the screen and that corresponded to each of the electrodes.

Biofeedback sessions would typically begin with the patient's breathing; an electrode would be strapped to the patient's midsection, and the therapist would ask her to use the lines on the screen to appreciate the difference between (less efficient) abdominal and (more efficient) diaphragmatic variations. Patients were instructed to practice the diaphragmatic style between appointments and to notice all of the bodily differences associated with it. Electrodes would then be moved to the jaw and inner thigh, and the therapist would gently touch the patient on different parts of her body (arm, ankle, face); then they would both watch the screen to see what kind of muscular tension registered in their jaw or inner thigh in reaction to this bodily contact. If there was a lot of tension—which was presumed to index a concordant amount of bodily fear or reluctance—the therapist might stay there for a session or two, before moving on to the vagina and pelvic floor. When it was time to involve the pelvic floor, it was monitored with an electrode that resembled a tampon; the therapist would, once again, touch the patient in random—and safe—places, and together they would watch the activity of the patient's pelvic floor muscles on the monitor.

The goal of these sessions was for patients to learn how to use their bodies (e.g., diaphragmatic breathing or a relaxed jaw) to produce lines on the screen that fell below certain numbers. The software in Cathy's office displayed a graph that rose to 5.0 microvolts, and patients were encouraged to work toward "2.0 and below," a level that indicated a relaxed (and theoretically pain-free pelvic floor. (A 1.0 was considered ideal, indicative of an uninjured or recovered muscular state.) Once this number was achieved, the patient would take note of the bodily maneuvers that she employed to get there, in addition to how her body felt at that number. Hanna would then encourage her patients to give this bodily state a word, one that the woman associated with calmness. (Daphne's, for example, was "Europe," based on a trip she had taken.) When the patient needed to call up that bodily state—for example, for dilator work, penetrative coitus, or a stressful day at work—she could shortcut the process with her word: "Body, go to 'Europe.'"

> CATHY: As that green line [on the monitor] starts to move in the lower direction, think "What am I doing to make that happen?" Now squeeze, tighten, and feel that. How does that feel different? [As the line drops, Cathy says:] Can you put words on that? You've dropped a full microvolt in a matter of minutes—can you put words [to it]?
> LIBBY: It feels . . . really loose.
> CATHY: Loose would be a good word. Now tighten and do the opposite of "loose." You might feel a difference, but that's an important difference. . . . My anus, my hips, my buttocks, my mind. Where do I sense that shift? Some people feel it in their feet.

Holding Patterns

One evening, when Libby didn't arrive for her scheduled 6:30 appointment, Cathy suggested, based on a previous discussion we'd had, that it might be a good time for me to try working with the monitor. Libby was her last patient, and Cathy was going to finish up some paperwork and close down the office—it was as good a time as any to take advantage of her equipment and knowledge. I agreed with only a slight hesitation, which related more to making myself vulnerable as a "patient" (and object of clinical scrutiny) than to any reluctance I had about exposing my genital body to Cathy. I quickly agreed to do it, however, and after plunking down thirty-five dollars for an individualized vaginal sensor (the nongenital electrodes were shared between patients), Cathy and I negotiated how much direction she would provide while I used it.

Patients were already undressed (from the waist down) when they did their pelvic floor biofeedback, as they had usually been engaged in "internal work" with their therapist beforehand. Because I was clothed, Cathy gave me the option (which I chose) of simply inserting the sensor into my vagina and letting the connecting cord trail out the waistband of my pants. This felt surprisingly stranger than being undressed, however, and I soon wished that I had opted for the more familiar experience of a drape sheet. I continued to chat with Cathy, however, optimistic that I would relax with practice and that I had an interesting source of data in front of me.

Since we knew this might be our only opportunity, Cathy and I cut right to the chase and didn't bother with jaw or breathing electrodes. We were interested in one line and one line only: the one that would tell us about the state of my pelvic floor. I will say that there were plenty of (self-reassuring) thoughts going through my head at the time, including many related to my earlier stories: I had grown very comfortable with this part of my body, I insisted that everyone I knew (including my professors) talk about it, I had had my fingers on or in the genitals of several friends in the past, and I would have been unconcerned had a friend asked me to examine her that very afternoon. I knew that, if called upon, I could have done a pelvic exam on Cathy without the slightest awkwardness, and I had long ago grown accustomed to talking with my own providers while they did mine. Moreover, I was fully expecting to see what Cathy herself was expecting (she told me later): a nice even 2.0, maybe even lower, reflecting my body's lack of genital or pelvic floor dis-ease. Indeed, part of how I coped with my tinge of embarrassment was by imagining the 1.0 that I was about to produce on the screen, a measurement that would have made me a pelvic floor hero, a body that could demonstrate what the right sensibilities, politics, and emotional investments were capable of.

But, as you have doubtlessly begun to suspect, that is not what we saw. Not by a long shot. Initially, my line was literally off the monitor. Cathy and I were both a little taken aback, and we chose to attribute my numbers to the newness of the situation. Cathy even generously suggested that they were due to the abruptness of plugging the electrode into the monitor and skipping over the breath and jaw work. So we just laughed a little and continued to watch the screen. Which didn't change much at all. We responded to our surprise by talking about it, and Cathy soon moved into her physical therapist role and began instructing me in techniques that I could use to manipulate the line. These were behaviors that I had heard her describe countless times, things I could have easily taught patients on my own. But that evening, I needed her to guide me; I felt awkward and like I was a disappointment—to her as well as to myself. Why couldn't I do this better? My line began to drop as I paid more attention to consciously relaxing, and within a few minutes it was hovering around a 3.0. Cathy, true to form, told me that she'd step outside and provide me with some privacy, a step I don't think either of us anticipated she would have to take with me. She left to begin closing up the office, and encouraged me to "play" (the word she used with patients) with my pelvic floor muscles and then watch them relax on the screen in front of me.

But left to my own devices, things did not get any better. As I sat in Cathy's recliner—sensor in place, the white cord creeping out of the waistband of my pants and attaching itself to the monitor, the machine that evidenced the hard facts of my personal "holding pattern"—I was both disappointed and relieved. I wanted my screen to look different, to be better. In the ways that I always wanted to get the best grade, I wanted to be an exceptional patient. But this was not just the kind of desire that comes of being from a large family or competition as survival skill—I wanted to be the best at *this* because this is what I *did*, what I had been training for throughout my entire feminist career. I wanted to believe that I could beat the machine, that my purposefully collected and "healthy" genital hexis was *enough*, that it could beat not just the biofeedback monitor, but, and more importantly, that it could beat what those lines represented—the disparaging discourses of inconsequence and shame that appear over and over again on these pages. And if I had done this, I could still attribute vulvar pain to a "cultural" or ideological condition in which the responsibility for symptoms would lie more completely with symptomatic women who just weren't *dealing with* their environments skillfully enough. It would be just like my early days as a clinician, where I simply needed to teach these patients what I knew to be true, what *I* had learned and done.

The lowest number that my monitor ever displayed that evening was somewhere between a 2.0 and a 3.0, and even that was a fleeting accomplishment. I was usually above a 3.0—whether alone or with Cathy, squeezing or releasing, laughing or tranquil. Cathy and I continued to awkwardly discuss what might be going on, such as my history of chronic urinary tract infections, which may have altered the behavior of my pelvic floor. I told Cathy that I had noticed myself doing more Kegel (strengthening) exercises during my fieldwork, as if the lack of control that patients had over their own pelvic floors was making me hyperaware of my own. But alongside my disappointment, I felt a small amount of validation regarding the research project in which I was engaged. I understood that I didn't beat the monitor because my body responds to disparaging genital discourses in the same ways that the other women represented in this book do; my pelvic floor is also a product and producer of unwanted genital experience. Cathy suggested that many women probably engage in some amount of regular vaginal and rectal tightening and then added, "For those of us who *don't have issues down there*, we can talk about this." When I suggested that maybe some of us might not know that we have "issues" until we (try to) talk about it, Cathy said, "Well, you're definitely holding *something*" (my emphases).

And I couldn't have agreed more. I *was* holding things, lots of them. Duden (1998) writes that our "notions of corporeality seem deeply embodied . . . like petrified deposits of the modern age to which we belong" (22). In my case, these deposits were constituted only somewhat differently than those of my informants. Without holding a muscular (and emotional) set of reactions to the exquisite tenderness of *vulvar vestibulitis syndrome*, or to the erosion of my vulva through *lichen planus*, my pelvic floor nevertheless held the deep-tissue discourses—the "background conditions of social interpretation and practice"—articulated by Povinelli (2006, 23). In telling stories of VVS and vulvodynia, I demonstrate how the experience of pain that is clinically superficial articulates and indexes the tenacious hold that vulvar dis-ease has over female bodies.

This book argues that a woman's relationship with her genitals should not be aversive, whether "it hurts down there" or not.

My fieldwork focused on women diagnosed with a pain syndrome, because the material reality of their symptoms forces them, in the most explicit ways, to confront the discursive vulvar shame that afflicts a far greater number of women. Faced with these contaminations and erasures, numerous women are capable of sending into hiding the parts of their bodies that are given—and simultaneously denied—meaning through disparaging discursive events.

Though not solely the property *of* bodies, diseases—like racialized or gendered identities—are typically identified *with* particular bodies through the accumulation of specific social histories. Patients in the midst of the diagnostic process cannot come to terms with their pain without acknowledging the material reality of their vulva. At the same time, they recognize and produce the circumscribed environment that informs the silence through which their pain has primarily been lived, a collective silence that transcends the singular experience of pain. Clinic patients acquire, and then cultivate, verbal and behavioral strategies that facilitate the medical legitimation of their symptoms. Physicians prepared to offer this validation meet patients across a dynamic threshold that begins to (re)organize their experience, and it is here where vulvar disease is most fully realized, where allegedly new symptoms find meaning and repositories, and where a growing number of knowledges, services, and experts proliferate.

It is not the presence of pain that has kept patients from developing a working knowledge of their genital anatomy; outside their diagnosis, they share in a dominant cultural vulvar reality that is marked by transgression. If we think in terms of cumulative layers, this shame-based self-reluctance is at the bottom, middle, and top of the genital aversion voiced not only by the patient I described earlier but by every clinic patient that has ever told her provider, "Well, I don't usually look down there." But in *realizing* their disease condition, symptomatic women have an opportunity to identify and create alternative layers. Some strategies already exist and need only be brought out of hiding, such as the ability (of many patients) to accurately pinpoint an area of pain no larger than a centimeter ("Um, it's usually right around here."); others are completely novel and require more substantial (and consistent) support, such as the purchase and regular use of vaginal dilators. The important point here is that these are behaviors that can be acquired by a wide range of bodies, not just clinicians with feminist politics or women with vulvar pain. Whether or not you have symptoms that meet the criteria for VVS, vulvodynia, or lichen planus, it helps to be able to say: "The pain is vulvar and seems to be aggravated by certain fabrics," rather than to simply answer "kind of" when (and if) your clinician asks if you have pain with intercourse. The genital familiarity and integrity that *can* accompany the diagnosis and management of vulvar pain is a corporeal orientation, a mode of comportment (Young 2005) from which a much greater population of diseased women might appreciably benefit.

The anthropological lens through which I engage with this pain is a critical factor in the development of such a collective shift. Left to their own devices, both diagnosed women and the physicians who care for them invest heavily in the physiological dimensions of these conditions, leaving newly acquired genital behaviors to be circumscribed by discourses of pathology and anomaly. In this framework, it is *only* women with pain that need greater vulvar awareness, and cultural vulvar dis-ease is effectively eclipsed by physiological disease. Asymptomatic women are not the recipients of increased genital attention, nor are the patients themselves if and when their pain is resolved. This was repeatedly evidenced by the number of diagnosed women who told me they couldn't wait until they didn't "have to deal with their vulva anymore."

In returning to my earlier theme of "baggage," then, I want to suggest that we think about the full range of (physical and cultural) genital behaviors with the broadest sense of this term. Many women carry the oversize and weighty pieces—the steamer trunks full of insults, erasures, and *dys*-appearances that constitute unwanted genital experience—whose unwieldy nature limits their bodily freedom. But it needn't be the case that only women with a recently recognized disease condition be given access to pieces of baggage that are lighter and more user friendly: uncontaminated self-examination, therapeutic attention, and vulvar integrity. In my model of dynamic layering, women can *choose* to accumulate and acquire the kind of genital "baggage" that facilitates alternative behaviors and bodily imaginaries, pieces that allow them to move with greater ease and flexibility. These may not—indeed, likely will not—replace the more familiar and overdetermined modes through which women carry their genital bodies. But in lightening the load of *in*consequence, they can begin to displace and reconfigure the bodily belongings that *any* woman might agree to bear.

4

MANIFESTATION

(Un)conscious Presencing

It's red. It's raw. I get these little cuts. I'm late for things a lot. It itches. It's irritated. It feels like sandpaper, like someone poured acid on me, like ground glass. It's stabbing. Knifelike. It feels like you're taking a knife to me. It just feels so vulnerable. It's going to hurt. I want to pull my knees in. My skin splits and tears. It's really sensitive. It feels like a razor cut. Like a wire of pain. I just tense up. It itches so much I just want to tear my skin apart. It's like there's a wall in there. A wall of pain. I want to just drag myself along the floor, pull the crotch right out of my pants. It's that one spot. It feels like someone hit me with a sledgehammer in my crotch. It burns. Like someone put lighter fluid up there and lit a match. Like I'm sitting in fire.

They sit with their legs crossed under them so that there is less contact between the chair and their bodies. They stand up after they've been sitting for longer than an hour and a half. They grip the sides of the examination table with fingers familiar with the soft give of the vinyl-covered cushion. They cry, but not as much as I thought they would. They are—and they look—desperate. They are hungry for information. They apologize to the doctors—for their symptoms, for their inability to correctly narrate their medical history, for being as upset as they are. They have scars, both from diagnostic biopsies and "corrective" surgeries. Their skin looks white where it should be pink, leathery where it should be smooth. They are "not themselves." They have lost the contour in the folds of their labia, the suppleness and mobility of their clitoral hood. They pull away from the touch of a hand, a speculum, or a Q-tip, with the speed and agility of a greyhound. Their bodies have a classic presentation of a disease.

—From my field notes

SHOWING UP

This chapter unpacks some of the ways that vulvar disease is realized in clinical spaces: the intimacy of a PT session, the clinical history of a patient, and a major medical conference. Through the concept of manifestation, I examine a range of cultural factors—behaviors, locations, actors, institutions, events, and structures—that coconstitute and contour the experience and ontological reality of vulvar pain (Mol 2002). I also explore how vulvar disease has commanded varied amounts of clinical, political, and cultural attention—how, in other words, it has shown up in the world thus far.

My analytical interest in this chapter is in troubling the superficial. It would be impossible for me to describe vulvar disease without taking note of how symptoms both appear and are perceived on a corporeal surface. As an anthropologist, however, my goals are distinct from those of the clinicians and researchers who work to subdue their patients' "skin pain." I work, instead, to animate the symptoms before me, to locate and expose the cultural deep tissue that manifests through surface pain. I think about the symptoms and bodily behaviors associated with vulvar disease as eruptions: material and observable realities that offer clues about the sociocultural processes that contour them. Vulvar specialists formulate treatment strategies designed to eradicate, contain, or redirect the physiological aspects of these eruptions, including pudendal nerve blocks, pharmacological manipulation of neurotransmitters, surgical excision of hypersensitive tissue, and the repeated application of topical anesthesia (Andrews 2011; Boardman et al. 2008; NIH 2012; Rapkin, McDonald, and Morgan 2008). By critically interrogating the medical manifestations of vulvar disease—including these commonly deployed treatments—I show how these bodily symptoms and medical behaviors operate within and reinforce accumulated discourses of silence and erasure. My attention to the amplified presence of vulvar pain takes exception to these received discourses, insisting on a fully contextual encounter with its troubling nature.

MANIFESTATION 1.
A VISIBLE FORM IN WHICH A DIVINE BEING, IDEA, OR PERSON IS BELIEVED TO BE REVEALED OR EXPRESSED.

Nikki and Lisa

I met Lisa when I accompanied Nikki to her first PT session. Nikki, whom I met on my first day in the clinic, was interested in having surgery; she had previously been diagnosed with VVS, and Dr. Erlich had agreed that she was a good candidate. Because the clinic physicians recommended PT for surgical as well as nonsurgical patients, Dr. Erlich referred Nikki to the in-house provider, a woman named Sandy who had—largely at the clinic's behest—begun to specialize in pelvic floor disorders.

Nikki and I quickly developed a comfortable friendship, one that came to include meals, coffee dates, and even a couple of yoga classes. We talked as easily about her history, her pain, and her marriage as we did about novels and fashion. Neither of us knew what to expect at her first PT visit, but our mutual ignorance worked to both our benefits, since what occurred that afternoon would have disrupted any notions we had of a typical session. Lisa was filling in for Sandy (who was on maternity leave), and though she did not share Sandy's level of expertise, she told me in a subsequent interview that she was keenly interested in genital pain: "This is my thing now," she said. I was to eventually learn that the two therapists embraced radically different approaches: Sandy's technique was rooted in correcting musculoskeletal misalignment and maladaptation, whereas Lisa worked from within an alternative bodily imaginary. Because Sandy's leave was temporary, Nikki worked with both therapists, leaving me to wonder whether her recovery would have proceeded differently had she worked exclusively with one or the other.[1]

Lisa began Nikki's sessions with some clinically focused questions, but she shifted gears quickly. As she assessed Nikki's body for the first time, Lisa asked her to describe how it felt to be touched in and around her genitals. When Nikki responded with "I think . . . ," Lisa interrupted her by gently commanding, "I don't want you to think. I want you to *feel*. . . . I want you to become aware of what's down here." ("I didn't know what to do," Nikki later admitted to me, but her overall reaction to Lisa's line of questioning was open and participatory.) Taking her cues from Nikki's receptivity, Lisa forged ahead by directly asking if there was anything that might be holding Nikki back from getting better, anything that would cause her to "hold on" to her pain: "Are you open to finding out that there might be some emotional/feeling work to do in this getting better? Because we're going to do some mind-body work in here."

It's the Body's Intelligence: The Mind-Body Connection

"Integrative medicine" is one term commonly used to describe medical practices informed by so-called alternative, or complementary, perspectives (e.g., acupuncture, massage, chiropractic, and herbal medicine), many of which purport to be more holistic than conventional biomedicine. My own thinking about vulvar pain had always been oriented around the mind-body connections invoked by these approaches, and I maintained that predisposition during fieldwork. Despite mounting evidence that vulvar pain was indeed physiological, it seemed shortsighted for physicians to wholly extract its emotional components from their diagnostic considerations. But though I knew that they, along with a cadre of dedicated researchers, were foregrounding the biomedical dimensions of vulvar pain in order to free affected women from vicious and unproductive cycles of self-blame, I was beginning to collect a number of nonclinical stories that were uncannily similar and that contained a surprising number of details about sexually reluctant, violent, or painfully prohibitive pasts—the *unwanted genital experience* theorized in this book. In my anthropological investigation of vulvar pain, a mind-body connection was taking increasingly substantive shape, fleshed out by the histories of women like Nikki who were more confused than anyone about how much significance to attribute to their individual and "emotional" pasts.

Pushing vulvar pain's affective dimensions into the background also had an economic component. Both Drs. Robichaud and Erlich told me, and would routinely tell their patients, that they "used to send all [the VHC] patients to Jill," the in-house sex therapist, but that budgetary constraints now prevented them from doing so. At one hundred and fifty dollars, Jill's hourly rate was prohibitive for the majority of clinic patients, making the decision to deprioritize emotional work both more understandable and more vexing. But Dr. Robichaud especially continued to impress upon her patients the salience of sexual counseling, and she did so in a way that left vulvar pain open to extraphysiological interpretive accounts. Whether or not she made a direct referral, Dr. Robichaud almost always told her patients about a study conducted by Sophie Bergeron, a Montreal-based psychologist who has been studying vulvar pain for close to two decades. In short, the study randomized women with vulvar pain to either PT, surgery, or counseling. No one in the sample received more than one intervention, and each group demonstrated statistically significant reductions in their pain (Bergeron et al. 2001). Dr. Robichaud used these data to explain that each treatment strategy had a measurable role to play in resolving vulvar pain; she further speculated that if the treatments were individually efficacious, then any combination would likely be even more so. Aside from the fact that her conclusion was (as yet) unsupported by empirical evidence, Dr. Robichaud's narrative was slippery for another reason: the evidence seemed to suggest that arguments for a *purely* physiological vulvar pain were at best incomplete and, at worst, specious.

These regular (and sincere) discussions of sex therapy both ruptured and adhered to the biomedical narrative espoused by the physicians and most of the PTs with whom my patient-informants worked; in other words, that the psychologically based elements of female genital pain were ancillary to its "real" and physiological dimensions. Counseling was seen to be important, even integral, to some women's recovery, but for those who could not afford it, or for whom therapy implied the kind of self-blame from which a medical diagnosis promised to free them, it was (fairly) easily excised from the treatment plan. But the nonphysiological aspects of these women's pain haunted my interviews, making the gap between what they might have needed and what they got that much more poignant. When I conducted extended interviews with four patients during one particularly busy week, for example, I was caught relatively off-guard when three of these women disclosed sexually traumatic histories that had not been discussed at their initial clinic visits. What was especially intriguing about this is that the clinic's official paperwork contained a direct question about such events, while my prepared interview questionnaire did not. This does not mean that the physicians were less concerned about these narratives; their intake questions signified their efforts to elicit such pasts.[2] Rather, the asymmetrical stories that we collected index the distinct valences accorded by each of us to pasts that were emotionally and culturally charged.

I left questions about sexually abusive backgrounds off my formal questionnaire for one major and historically specific reason: I wanted to dissociate myself from accounts of vulvar pain in which sexual trauma was the implicitly identified culprit. In the absence of clinical *signs* that, by definition, are the measurable and observable complement to subjectively experienced *symptoms*, uneducated clinicians have often taken diagnostic refuge in the category of psychosomatic illness (Bodden-Heidrich et al. 1999; Pucheu 1998). This clinical dilemma was further informed by two significant historical developments: a post-Freudian climate in which even physicians without psychoanalytic inclinations remained unavoidably aware that inexplicable symptoms might have a traumatic foundation (Harlow and Stewart 2005; Raphael, Widom, and Lange 2001) and a feminist-oriented sexual assault movement that effectively convinced survivors, advocates, and care providers that trauma could be embodied (Brownmiller 1976; Cvetkovich 2003; Herman 1992; Ruch, Chandler, and Harter 1980; Winkler and Wininger 1994). In this context, linking inscrutable (and penetratively prohibitive) genital pain with allegedly buried bodily memories of sexual abuse not only made sense, it did so on multiple levels. Indeed, I sometimes cringe at the memories of my own professional encounters with these symptoms,[3] during which I routinely defaulted to so-called psychosocial explanations. Once I had determined that I couldn't find anything "wrong," I—like the rest of my colleagues—almost always referred these patients to counseling.

But however attuned to embodied trauma these 1980s-era clinicians might have been, our routine referrals and deferrals provoked a grassroots backlash—an eruption—among some of the most resourceful women with symptoms.[4] Tapping a collective nerve, the work of these early activists focused on unsettling and disrupting a connection that, though sensible in theory, was perceived to be devastatingly ineffective in practice. Catalyzed by many of the same feminist health movement principles through which their pain was thought to make psychological sense, many symptomatic women resisted the framing of their pain as nonphysiological and began to press for further—and federally funded—biomedical research. Fortunately for them, their demands spilled into an arena that was prepared to listen, and, in a matter of years, clinical research and review articles about vulvar conditions shifted substantially in their analytic frameworks and orientations. A somewhat random, and often puzzled, series of hypothetical descriptions of the psychological and "character" differences between women with and without symptoms transitioned into a more systematic stream of demographic and correlational analyses. This latter group of studies sought to statistically demonstrate relationships between the presence of vulvar pain and a host of other recordable factors, including age at first coitus, menstrual patterns, and concomitant infection with sexually transmitted viruses such as HPV.

It was these early activists, and their symptomatic descendants, with whom I was to talk and conduct my research. Because the 2000s-era wave of research repeatedly—and numerically—"failed to find an association" between vulvar pain and a history of sexual trauma (Harlow and Stewart 2005, 871; see also Dalton et al. 2002 and Edwards et al. 1997), I was cautious about evoking or imposing this controversial variable; indeed, my intention was to leave it out of my investigation unless it emerged on its own. When it did, it was with a frequency and quality that neither reimposed a purely psychosomatic rendering of vulvar pain nor left me in a conceptual lurch regarding the broader mind-body connections that I was reluctant to surrender. In other words, the number of women that revealed histories of a traumatic nature was consistent with the vast majority of published prevalence rates (Dalton et al. 2002; Edwards et al. 1997; Cvetkovich 2003), most of which estimate that close to 20 percent of US women will be sexually victimized in their lifetimes (CDC 2010; RAINN 2012). Although such rates often serve varying agendas, their consistently high numbers help to substantiate the claim that symptomatic women are, in fact, "no *more* likely" (Harlow 2004, personal communication) to have survived some kind of unwanted sexual experience: reports of sexual abuse are high in *every* population, period. In my analysis, this does not obviate the significance of these experiences. Rather, the discrepancies between the respective histories gathered by the physicians and me are what allowed me to consider the term "unwanted genital experience" in a far more expansive context.

In the clinic, patients were asked to check a box marked "Yes," "Some," or "No," next to the statement "I have been sexually abused." But these written responses did not always translate into an explicit discussion of the relationship between past experiences and present symptoms. During the visits that I observed, these histories were typically not discussed unless initiated by the patient. And indeed, posing the question in this way risked missing several groups of women: those who chose not to disclose at all, those who preferred to disclose verbally, and those who did not recall or interpret their unwanted genital experience in "abusive" terms. Conversely, my interview questions left the direct version of the question unasked but purposefully opened up a space through which a multiplicity of experiences—bodily, psychological, *and* cultural—could be apprehended as genitally unwanted or deleterious. Moreover, and cognizant of Ann Cvetkovich's (2003) call to celebrate, rather than disparage, some of the established sequelae of incest and sexual abuse, such as lesbianism and queer sexual scripts (89–92), I was keen to leave my informants' narratives as *under*determined as I could.

Through connecting vulvar dots that were arranged in patterns both erratic and predictable, my informants and I asked questions about how routinized social censorship articulated with symptoms that went undiagnosed for five years or longer. We wondered together about why it was so hard for them to talk about their pain with even their closest friends, and why their husbands never talked about it with anyone. We often described conflicted and contradictory attitudes toward sex, orgasms, and pleasure, and patients repeatedly stressed that their sincere desires to fiercely enjoy sex with their husbands stood in stark contrast to the religious (or otherwise sexually proscriptive) values with which they were raised (Carpenter 2005). In our interviews, patients and I redefined the word "unwanted" by considering the relationships between their various life histories and their symptoms as not necessarily causal, but neither as insignificant nor somehow ancillary.

Safety and Support

As I was wrapping up my fieldwork, the Harvard researchers (Harlow and Stewart) whose 2003 study I have previously discussed published a second article in which they revisited some of the facts beginning to congeal around vulvar pain risk narratives. Having already learned that a substantial number of symptomatic women were not showing up in clinic spaces (particularly nonwhite women), these researchers questioned the "no relationship between abuse and pain" data on the grounds that it had been collected only in these specialty clinics. And because they had honed an interview script that effectively screened *in* clinically verifiable disease (Harlow and Stewart 2003), they pursued the same strategy, seeking, in their words, "to determine whether fear of or actual childhood victimization, including sexual and physical abuse, influenced the risk of vulvodynia" (Harlow and Stewart 2005, 871).

Not surprisingly, they found a number of associations between burdened pasts and painful presents, the most striking of which was a fourteen-fold risk of *vulvodynia* for women whose histories included the "joint effect of severe abuse, childhood endangerment, and lack of family support" (2005, 875).[5] Disentangling these variables produced risk profiles that were somewhat less onerous but that still described a two- to fourfold risk for women who felt endangered at home or school, who "never or only rarely received family support as children," who both feared and experienced "severe physical abuse," or who experienced sexual abuse at the hands of a distant relative or friend (the risk rose to sixfold if the sexual abuse was perpetrated by a "close relative") (874–75). Based solely on these numbers, and given the nature of her childhood, Nikki's vulvar pain was almost predictable; the unwanted experience of her past (as I discuss later) involved not her own genitals but the safety and integrity of her mother and sister at the hands of a physically and sexually abusive husband and father.

Unpacking these histories, that is, those of my informants and that of vulvar pain research, is important if we are to examine links between epidemiological data like Harlow and Stewart's and several of the other concepts central to my broader argument—unwanted genital experience, safety, alienation, integration, and the bodily *home* made vulnerable by *genital dis-ease*. Establishing such links is critical in order to demonstrate the institutional dimensions of Vulvar Disease—how it "shows up" in bodies and practices that are informed by trauma, an experience that Cvetkovich (2003) reminds us was initially defined by Freud as the penetration or "breach" of a protective shield (52–56).

Harlow, Stewart, and I agree that the emotionally unsafe backgrounds of vulvar pain patients are salient; indeed, they conclude their 2005 study by stating that "to understand the etiology of vulvodynia, more emphasis should be directed toward assessing early-life exposures" (879). But we disagree about constraining "safety and support" to a set of *individually* inflected variables that were not only previously rejected by both patients and advocates but that also do not address the broader cultural context through which US women live their genital pain. In psychological terms alone, women have higher rates of PTSD (Herman 1992; Olff et al. 2007), and they routinely score lower than men on measures of well-being (Frank 2000; Piccinelli and Wilkinson 2000; Silverstein

2002); moreover, lesbian women may be at even higher risk for psychological trauma (Szymanski and Balsam 2010). Thus, though I applaud Harlow and Stewart's almost singular efforts to reentangle vulvar pain with past experience, as well as to make room in their research agenda for the measurable and physiological effects of onerous childhoods (Gerber et al. 2002; Harlow and Stewart 2005, 879), I find their conclusions about vulvar pain's ultimately "elusive" (879) nature to be stubbornly uninformed by the insights that anthropology and "feminist curiosity" (Enloe 2004, 561; Zimmer 2007) can provide.

Veronica

In her second session with Nikki, Lisa introduced craniosacral therapy (CST), a practice developed by an osteopathic physician that involves subtle manipulations of the vertebral column and cranial bones in order to balance the central nervous system. CST practitioners believe that bodily trauma can lead to blockages of the cerebrospinal fluid, and that gentle work with the spine and skull can ease restrictions of neural passages, optimize the movement of cerebrospinal fluid through the spinal cord, and restore misaligned bones to their proper positions. CST is closely aligned with so-called complementary health care, and patients are often converted to its logic through traumatic events that have been both life changing and ineffectively addressed by conventional medicine.[6]

Lisa requested Nikki's cooperation in moving to this technique and, after Nikki assented, added the caveat that they "wouldn't do any dialoguing [that] day" due to being short on time. When Nikki asked, "What's that?" Lisa said:

LISA: You can ask parts of your body questions.
NIKKI: Who answers?
LISA: You'd be surprised. . . . We do a lot of [dialoguing]. Because a lot of people come in here, and the doctor doesn't know what's wrong. And guess who knows?
NIKKI: The person?

LISA: The body. It's *the body's intelligence*. [Lisa says she's a kind of "mediator" between Nikki and her body.] A lot of times, the pain is a mediator, considering how the doctors have done all the physical things they can do. So now it's time to see if the body has some other stuff going on.
NIKKI: *Fear*. It's the only thing I can identify. [My emphasis.]

In this exchange, Nikki connects her *VVS* pain to a "fear" that her body "knows." Conventional understandings of embodied trauma, as well as psychoanalytic techniques, would use Nikki's identification of fear as a way to displace her pain from her body to her mind. Clinical vulvar specialists make room for the expression of such emotions, while simultaneously disavowing their etiological role in the development of vulvar pain. The VHC physicians (and many conventional PTs) believed PT's success to be based in the physiological changes that manipulating tension-filled muscles could produce (Gentilcore-Saulnier et al. 2010; Goldfinger et al. 2009; Hartmann 2010). In this model, addressing Nikki's fear would be akin to the "reassurance, explanation, [and] advice" that many clinicians suggest should accompany a full "exploration of psychological issues" and "a good doctor-patient relationship" (Talley 2001, 2062). Such interventions are based firmly in an understanding of physical symptoms as *indirect* effects of psychological (or cultural) events and processes.

In their practices, however, therapists like Lisa identified somatic locations for feelings of guilt, confusion, fear, and even joy in the bodies of their patients. Their interventions dialogued with a body capable of feeling, knowing, and responding directly to its social environment. This body, as Harlow and Stewart (2005) have also now argued, can develop a pain syndrome under the influence of what they refer to as "childhood victimization" (878–79) and what I refer to as unwanted genital experience: a pain syndrome *unmediated by other psychological states* (e.g., depression) and therefore not psychogenic (see also Raphael, Widom, and Lange 2001).

Feminist and subaltern theorists have identified the importance that naming has in the construction of a new social reality (Moses 2012; Murphy 1991; Rushin 1989), and vulvar pain syndromes cannot exist—indeed will never be "real"—without a recognized set of linguistic clinical constructions. Lisa's work with Nikki took this process a step further when she used their third session to speak directly with Nikki's body, to ask it—as did Freud in his earliest sessions—to "get in on the conversation" (Wilson 2004, 10). But before doing so, Lisa asked Nikki to give her pain a name. "Are you up for that? Are you into that?" she asked. Nikki, who had been "putting off" addressing her pain for almost three years, readily complied and named her pain Veronica. Lisa and Veronica spoke easily, and, after only a few sessions, Lisa asked Veronica if she would consider "stepping aside" so that Nikki could enjoy a sexual relationship with her husband.

> LISA: Veronica, Nikki is in my office. . . . She's an adult now and she wants to have sexual . . . relations. You have been there for her, you've been there for her in the strongest sense of the word. Can you recognize that she's an adult?
> NIKKI/VERONICA: I don't know.
> LISA: Can you trust her that she can recognize what's safe to put into her body?

Here, Lisa is attempting to uncover some of what motivates or surrounds Nikki's painful vestibule. Earlier, Lisa told Nikki (of her pain), "It's protecting you," to which Nikki replied, "Yeah. From pain."

> LISA: Veronica, when did you start protecting Nikki? When she got married? Before that?
> NIKKI/VERONICA: When her parents would fight.
> LISA: When was the first time?
> NIKKI/VERONICA: In Marshview.[7] [Nikki then relates a memory of her mom on the phone with her dad, crying over a conflict they were having. Nikki had

never seen that before, and she recounts feeling like she needed to protect her.]

What is crucial to notice in these dialogues is that Nikki contends that Veronica, in other words, her genital pain, is protecting her "from pain," seemingly of another nature. In one of our many conversations, Nikki told me that her father was emotionally, sexually, and physically abusive to her mom and sister, but that she herself had been exposed only to his emotional abuse. Nikki also recounted being the one who tried to "stand up for" and "protect" the other women in her family and that she often felt guilty for being spared by her father. When she was a teenager, Nikki's dad violently committed suicide, an event with which she had since made considerable peace. She shared these stories with me while we lingered in the waiting room one morning; she spoke about the guilt with which she still lived, and we wondered together if her genital pain wasn't some way of managing these feelings. By disallowing herself any sexual pleasure, Nikki could atone for the disparate and horrific treatment suffered by her mother and sister and still maintain a peaceful acceptance of her father's violent and painful life and death. This was further enabled by the fact that her mom was now happily remarried and was close with Nikki. "She's amazing," she said of her.

Importantly, Nikki had been making connections between her painful past and present long before PT. This was evidenced during our formal interview weeks before her sessions with Lisa when, seconds after I pressed the red button on my tape recorder, Nikki volunteered, "It's important to note that, growing up, my parents were incredibly dysfunctional." She continued that her mom had "found her voice" in counseling, however, and had subsequently taught Nikki "how to say no and set boundaries with [her] dad." The model provided by her parents for an intimate/sexual relationship was an unfortunate one for Nikki; for as long as she could remember, she knew that her mom "detested [sex], hated it, [and] couldn't say anything nice about it."

She says that "the most negative thing for me was growing up." She then told me a story about driving in the car with her mom and sister one day. Her wedding night "appeared" to her, "the night you would consummate." She asked her mom about it and her mom's reply was negatively inflected. She says she doesn't blame her mother: "living with a husband that, um, where rape is involved? I can't blame her for that. I can't blame her for anything that she feels, you know. 'Cause I, I can't imagine what she . . . went through, and I don't want to imagine, you know. So . . ." (Field notes)

I mentioned to Nikki that I found it interesting that she began our interview by invoking her "dysfunctional" family, and when I asked her how (and if) she thought this was connected to her pain conditions, she replied:

NIKKI: Mmm. I think it's really hard for me to feel . . . um . . . I guess *positive* about it. And it's taken me a lot of years . . . of . . . continually telling myself, "It's okay. You know, it's okay to . . . have sex."
CHRISTINE: Mmm-hmm. And to like it.
NIKKI: Yeah! And to like it, and to . . . want it more than, you know, whatever, it just . . . Yeah, that was *really really hard* at the first of our marriage as well. Because I was like "Oh gosh! This is kinda bad, and . . . and I didn't say that! And I told Sage that. I mean, he knew, you know, he knew the dynamics of my parents and stuff. And I said, you know, it ju— It really sucks because I know that it's okay, and I've waited this long, and I've done everything that's supposed to be right, and yet I still feel *guilty*. Yeah, that was, that was a lot of it. Guilt at the beginning.
CHRISTINE: [Do you] connect the guilt with the pain?
NIKKI: [Long pause.] I think the guilt was something I had to work through emotionally. Um, in respect to, I guess just the touching between my husband and I?

And ... I was able to ... not think of it as such a bad thing. But the pain was still there, and so to me they're kind of like two separate issues. And at the beginning I didn't know what it was.

Harlow and Stewart encourage us to take Nikki's entire story into account and allow for the fact that her "classic presentation" of VVS includes extraphysiological factors. Aside from these two researchers, however, such pasts have been increasingly marginalized as clinical research agendas have zeroed in on the physiological dimensions of vulvar pain (NIH 2012). In using Nikki's understanding of fear and pain that, in her case, were related to a violent dynamic in her family, I am calling attention to the "other causes" (Harlow and Stewart 2003, 83) of genital pain invoked in my introduction to this volume. Nikki's narrative confounds a physiological focus in that she cannot easily distinguish her "skin pain" from the pain of her parents' unhappy and violent marriage; she locates Veronica's (her pain) arrival squarely in the midst of a conflict between her mother and her father. Nor can Nikki make ready distinctions between the "guilt" she feels about her inability to protect the women in her family and her "guilt" over enjoying sex. This latter guilt is compounded by not only her mother's troubling sexual history but also by the Christian theological narrative with which she was raised and through which premarital sex was proscribed. Indeed, Nikki recalls that Veronica arrived when her family was living on the coast, the only time (she told me) when both of her parents were "active in the Church."[8]

My anthropological attention to alternative "etiological pathways" (Harlow and Stewart 2003, 87) for vulvar pain is an attempt to establish a safe and supported place for understandings of these conditions that more closely resemble Lisa's. Unwanted genital experience arrives in packages that are deeply—sometimes horrifically—personal, as well as in collectively experienced social structures. Though uniquely hers, Nikki's story is informed by widely prevalent patterns of gendered violence and sexual subordination, experienced by women who routinely prepare for the disparagement, harm, or disappearance of their genital bodies (Bartky 1990; Hlavka 2014; Solnit 2014). In telling me "Whether I was *hot* on sex, or whether I hated it, I would still have [this] pain," Nikki concluded an intense and intricate narrative with a neat separation between her *feelings* about sex and her ability to *have* it. But her story includes factors that blur this division: she has *both* a vulvar vestibule that may (or may not) overproduce inflammatory chemicals *and* feelings about being sexual that grew, in part, from the gendered vulnerability and hypervigilance under consideration in this book. In her study of migraine pain, Kempner (2014) points out that this kind of compartmentalization often serves a useful purpose, particularly for patients whose conditions are socially stigmatized; drawing distinctions between *me* and *my pain*, she argues, "is a way of granting distance from" and perhaps even "assert[ing] some control over" one's symptoms (100).

Kempner concludes that there are risks to this strategy, however, as it may contribute to revised forms of biological reductionism through which women's physical vulnerabilities remain stubbornly attached to narratives of inferiority and weakness, a point with which I concur. A useful approach to this dilemma, that is, analyzing disease conditions that are more likely to afflict subordinated populations, comes from scholars like Anne Fausto-Sterling (2004), who, in the case of hypertension among African Americans, urges us to see "*an orchestrated response to a predicted need to remain vigilant to a variety of insults and danger*—be they racial hostility, enraging acts of discrimination, or living in the shadow of violence" (26; my emphasis). Like these populations, women in the contemporary United States live their "biological" symptoms in and through a social milieu; the states of alienation and inconsequence that come with having a vulva—the "'accumulated insults' of living in a [sexist] society" (Pollock 2012, 102)[9]—should not be omitted from analyses of vulvar pain.

It is possible that Veronica is a liberated manifestation of Nikki's personally traumatic past, a restored connection now available for her recollection. Or perhaps she is a representative—a divine being—of the psychic pain shared between *any* women whose genitals have gone into hiding. But we need not explain why or how she showed up for Nikki in order to apprehend the intimate relationship that she has with the unwanted genital experience examined in this book. Grosz has argued that "the body functions not simply as a biological entity but as a psychical, lived relation" (1994, 27). In outlining some of the manifestations of women's lived relationships with a disparaging world—somatic modes of attention (Csordas 1993) that are particularly female as well as historically specific—I am expanding Grosz's assertion so that feminists will not limit their understandings of unwanted genital experience to realms more individual than collective. In revealing layers of dis-eased cultural injury—the "deep-tissue discourses" considered in these pages—might it be possible to imagine alternative and multiple pathways toward the resolution of Vulvar Disease?

ERUPTIONS

As a "style of bodily comportment that is typical of feminine existence" (Young 2005, 31), the genital alienation that was and remains the impetus for my research is in a dialectical relationship with a world that "is felt and functions as an extension of [the] body" (Salamon 2010, 60). The accumulated layers described in the previous chapter constitute an embodied predisposition through which the experiences of pain and unfamiliar symptoms are then organized and produced, for both unaffected and symptomatic women. What manifests and erupts—on the skin, in the clinic, during sex, and on the tongue—is obliged to this sedimented silence.

The problems associated with vulvar pain (physical symptoms, marital discord, clinical maltreatment) reveal the frayed edges of threadbare cultural practices whose utility has come into question. Pain and redness erupt onto the skin of affected women, signaling for some a multisystem outbreak. Like canaries in the proverbial coal mine, women whose bodies refuse heteronormative penetration and the tools of medical gynecology represent a social irritant, a raw and sensitive rash that is both increasingly legitimate and stubbornly intractable. In Freud's words, "They would not have become symptoms if they had not *forced their way* into consciousness" (1917a, 345; my emphasis).

In her 2004 essay "The Brain in the Gut," Elizabeth Wilson interrogates the level of "distance" (39) maintained by gastrointestinal physician-researchers from the emotional aspects of conditions such as irritable bowel syndrome and Crohn's disease,[10] a distance strikingly similar to the one maintained by my clinician informants. Wilson outlines contemporary neurological accounts of an enteric nervous system, which she defines as a "complex network of nerves that encases and innervates the digestive tract . . . [and that] may act independently of" (34) the brain and spinal cord; her goal in doing so is to "turn . . . our attention to how . . . distal parts of the body . . . have the capacity for psychological action" *in and of themselves* (34). In other words, the "emotional" components of a disease located in the gut, specifically one that is distinctly and physiologically innervated, may not *necessarily* involve the interpretive work of the central nervous system.

My interest in Wilson's "brain in the gut" stems not from my belief in a genital nervous system through which I can interpret my findings, but rather in how she reconciles Freud's earlier work on the nervous system with the paradigm shift in gastroenterology. Wilson recuperates Freud's work in several of the essays in her book *Psychosomatic*, but this essay in particular explicitly demonstrates the integrated nature of Freud's prepsychoanalytic therapeutic regime, an approach that uncannily resembles that of my informants who were physical therapists. Frau Emmy, a patient whose hysteria manifested through a range of gastrointestinal symptoms, provides the basis for this reconciliation.

In brief, Freud came to understand Frau Emmy's anorexia and digestive troubles as hysterical somatizations of a "number of [earlier] disgusting episodes concerning food and drink." Freud's work with Frau Emmy involved using hypnosis to clear her disgust and to establish connections between her gastric pain and her broader experiences of fear and anxiety, some of which began after her husband's death. After Freud administers an eclectic set of interventions, including "stroking her a few times" across the abdomen, Frau Emmy is cured, measured—in

Freud's account—by her ability to eat and drink easily on the day after his final treatment. Wilson's recounting of this case is not aimed at convincing us of the efficacy or innovation of Freud's techniques; rather, she is interested in Frau Emmy because she provides a "starting point for thinking about . . . how a husband's death, a patient's resistances and fears, and an analyst's authority can be gastrically internal—not just ideational or cerebral" (33).

Wilson's analysis of Emmy is meticulous, and her conclusions are prescient regarding the reality of Vulvar Disease. How, in other words, other than *genitally internal*, should we describe a pelvic floor that has hardened into a mass of impenetrable knots? How should we understand skin so hypersensitive that the strength of this same pelvic floor helps it to lurch away from encounters with cultural and anatomical phalluses: Q-tips, therapeutic fingers, and, in the words of my informant Mira, "even a cheese doodle"? The narratives in this book allow us to consider Vulvar Disease as not only embodied fact but also as cultural assemblage. If the body is the "existential ground of culture" (Csordas 1994, 135), then what is it that manifests when we critically juxtapose the "ontological and relational complexity" (Wilson 2004, 20) of a sexually repressive background, the impact of being called a cunt, the proliferation of cosmetic labiaplasty, and genital pain that is explained (by some) as the result of "fired up" nerves?

"Woman," according to Cixous and Clément (1975), "has always functioned 'within' man's discourse . . . [in an] energy [that] puts down or stifles [her] very sounds . . ." (95). From my desk in Blacksburg, Virginia—years and miles away from my encounters with these resilient women—I want to assign their pain a great meaning. I want to suggest that its excessive and confounding nature is nothing less than the genital manifestation of a female uprising demanded by these feminists four decades ago:

> It is time for her to displace this "within," explode it, overturn it, grab it, make it hers, take it in, take it into her wom[a]n's mouth, bite its tongue with her wom[a]n's teeth, make up her own tongue to get inside

of it. And you will see how easily she will well up, from this "within" where she was hidden and dormant, to the lips where her foams will overflow. (Cixous and Clément 1975, 95–96)

The dermatological, neurological, and inflammatory eruptions that constitute clinical vulvar disease provide both a vehicle and a forum for women to incarnate their genital bodies; indeed, their disease cannot be medically confronted without new corporeal behaviors. Whether these imaginaries perform the incandescent act of displacement theorized by Cixous and Clément is a question with which my ethnographic analysis is engaged. Duden (1998) argues that subjectively experienced "burdens and trials"—along with the meanings attached to them—were effectively excised from scientific medicine's modern and "objective" approach to the suffering body (30). By ascribing a meaning to Vulvar Disease that both accounts for and transcends *individual* psychic events or traumas, I situate patients and providers within a reconfigured medical milieu, one that has as much room for empirically observable markers of disease as it does for the residue and deposits of a dis-ease for which standard instruments of measurement have not (yet) been developed.

Three Millimeters

If I had to conjure an icon for my concept of manifestation, it would be Judy. I didn't meet Judy at her first appointment; rather, I heard about her while I sat in the pod with Dr. Erlich one morning, gathering fact sheets for the patient she had just seen. Dr. Robichaud and a new resident appeared in a white-coated blur—animatedly conferring, yanking various forms from file cabinets and hastily scribbling on them, and getting on the phone to arrange an obviously urgent surgery for the patient they had just seen. Dr. Robichaud told us that their patient had one of the severest cases of *lichen planus* (*LP*) she'd ever seen—her labia were so fused together, she was urinating through a three-millimeter vulvar opening. The procedure they were trying to arrange would surgically correct the problem, as well as evaluate how much overall vaginal patency it was possible to restore. After Dr. Robichaud finished sketching out these details, Dr. Erlich, who did regular volunteer OB/GYN work in parts of Africa, sighed and said, "Wow, it's like she's an Ethiopian woman." Dr. Robichaud, without missing a beat, said simply, "Yeah. It's like she's been infibulated."[11]

As I have described, LP is an autoimmune condition involving an overproduction of inflammatory discharge in the vagina. If not halted, the discharge's inflammatory nature contributes to permanent scarring, along with compromised patency and elasticity of the vagina. Anatomical proximity and gravitational pull lead to concomitant and deleterious vulvar effects; these include contour erosion and a loss of suppleness of the labia, as well as decreased flexibility and mobility around the clitoris and its hood. The decreased vaginal patency thwarts penetrative efforts—a notably functional complement to the otherwise cosmetic issue of vulvar contour change and loss.

In Judy's case, this form-versus-function distinction had been rendered moot by the severity of her symptoms; she had not noticed her labia's decreased elasticity (they were literally fused together) until she could not urinate normally. Unlike Mary Hudson (see chapter 2), for whom "everything still w[orked]" at the time of her *lichen sclerosus* diagnosis, Judy's profound loss of function forced her to take notice of skin changes that many affected women do not. Her clinical presentation served to remind her that an unobstructed genital opening (introitus) indexed a greater number of bodily possibilities than simple penetration. A three-millimeter vulvar opening not only occludes vaginal entry, it severely circumscribes what can exit the genital body, hence Drs. Erlich and Robichaud's likening of Judy to an infibulated woman. Judy herself could tell the doctors that it "was taking [her] ten minutes to pee," but she remained unaware of the role that her fused labia played in compounding her disease condition. The miniscule orifice that she presented in the clinic that morning did not allow for adequate expression of the vaginal discharge that was at the heart of her symptoms.

Judy's story—and the irreversible skin changes and loss that her genital body had sustained—made me incredibly sad. When I came home from the clinic that day and recounted the details to my housemate (a man in his early thirties who always listened to my stories with genuine curiosity), he asked me why and, more precisely, "how" this could happen to a woman with health insurance in the contemporary United States. To which I responded: "Because nobody gives a shit about the genitals of a sixty-two-year-old woman." I use this next section to further this assertion, underscoring that Judy's age only compounds (rather than causes) the discourses of inconsequence through which her sexual body is interpellated. I demonstrate how the three-millimeter opening in Judy's vulva indexes the vastly insufficient perspectives of the providers previously charged with her care. The compromised access that Judy and the clinic physicians had to her (sexual) body resonate—both materially and discursively—with the inadequate capacity of these nonexpert providers to properly attend to her disease condition.

Genital Preservation

Autoimmune diseases are notoriously enigmatic, and clinical presentations often elude exact diagnostic categories and treatment regimens. LP, however, is fairly easy to recognize and manage by knowledgeable gynecologists and dermatologists. Autoimmune conditions have long been treated, particularly in their more acute and life-threatening presentations, with systemic or locally applied steroids. This approach is a kind of catch-22, however, in that suppressing the inappropriately active immune system compromises its ability to ward off other potential pathogens. Physicians working with these conditions have begun using immune system modulators—which were developed to mitigate the physiological rejection of transplanted organs—precisely because they do not work through suppression; this generally means that they can be used with considerably less caution in patients with otherwise healthy immune systems.[12] Prescribing these drugs off-label for LP and

LS minimizes both the amount of medication that patients need to use as well as the number of side effects they are likely to experience.[13]

It is important to grasp some of the details of these drugs and of their use by expert providers, however, if we are to adequately analyze how Judy's vulva had been rendered invisible prior to her consult with the VHC. Judy had a master's degree in nutritional science and had worked in research hospitals in another part of the country for many years; many clinicians would call her a "medically savvy" patient. When Judy first began to notice her symptoms, she not only examined herself physically, she quickly sought the advice of a dermatologist friend. Although in some ways this friend's assistance was the beginning of a misguided series of interventions that led to Judy's severe clinical presentation, Judy was happy to have solved the problem. He diagnosed her correctly over the phone, based on her description, and she quickly filled the prescription for the topical steroid he prescribed. Since this was several years before the somewhat routine—and, again, off-label—use of immune system modulators, her physician friend cautioned Judy to use the medication conservatively, that is, to back off when her symptoms were under control.

Judy complied with her friend's recommendation. But it was not long before the steroid could not control her symptoms, and her friend referred her to a dermatological colleague in Judy's area. In the hands of this expert, Judy was biopsied and given a definitive diagnosis of LP; she was also switched to a higher-potency steroid and an immune system modulator called tacrolimus. Again, despite the decreased risks associated with tacrolimus, Judy was encouraged to use the medications only when her symptoms were troublesome. The problem with this regimen, however, is that LP is an unpredictable and idiosyncratic condition, equally likely to flare in stressful and stress-free situations. For this reason, physicians like Drs. Erlich and Robichaud encourage their patients to use the medications liberally and regularly at first, in order to establish good symptom control; subsequent backing off is done under the guidance of the doctor and in order to establish whether particular stressors can be identified, predicted, and avoided. This seemingly small material difference—the amount of medication prescribed by the physician—is in part a reflection of the relatively conservative nature of a provider's clinical orientation, a dynamic encountered by patients of all stripes, particularly those seeking new or off-label treatment regimens. But in Judy's case, the amount of medication prescribed by Drs. Robichaud and Erlich indexes a distinct orientation toward the relevance of the vulvar body, one invested in not only its anatomical and discursive *presence* but also its well-being.

Under the care of her more conservative dermatologist, Judy's LP (which she still poorly understood) became so severe that her labia fused together in the same way they had when she presented in the VHC. Although not as clinically urgent, in the sense that she could urinate normally and her vaginal opening was technically patent, Judy's labia were markedly flattened in contour and she could not accommodate any vaginal penetration. Significantly, Judy and her husband were having "difficulties" at the time, and their sexual activity had more or less ceased. Judy shared this with her physician, who subsequently recommended that Judy just "leave [her vagina] closed" unless and until she "needed it" again. Unaware of any other options and in a relationship with her genitals that was also penetratively circumscribed, Judy agreed to the plan. It was just over a year later, when the urinary problems described here began, that her dermatologist referred Judy to the vulvar clinic, aware that she *now* needed corrective surgery.

The *now* of this physician's decision adds another layer to the differences in kind (rather than degree) between the providers through which Judy came to understand her condition and my physician informants. At the VHC, new LP patients were not only encouraged to use liberal amounts of both steroids and immune system modulators in order to achieve good symptom control, they were also taught to understand the nature of their affliction. Neither medication will stop the (over)production of LP's vaginal discharge; in fact, life stressors and other factors will more than likely lead to occasional exacerbations, even with good pharmaceutical control, making every LP patient at risk for labial contour change, erosion, and vaginal scarring. In extreme cases like Judy's, the vagina fuses together unevenly, leading to an "apple-core" shape that the clinical literature describes as a classic presentation. Surgery is recommended to cut through the fused middle area and restore so-called normal vaginal patency.

As vulvar experts (as well as surgeons), both Drs. Robichaud and Erlich take this type of advanced presentation into account when they encounter new patients, and their efforts are actively informed by their interest in preventing it. Although I attribute this practice to their distinct orientation toward vulvar well-being, it is also true that it constitutes good preventive medicine. Managing a patient's chronic (or acute) condition as if it could worsen at any time is standard clinical practice in any specialty area, and most providers routinely do this with a wide variety of diseases (e.g., diabetes, hypertension). In this larger health care context, managing LP *without* taking this complication into account is at least correlated with (if not guided by) a *dis*investment in the preservation of a symptomatic woman's genitalia.

Since LP typically afflicts women in their postreproductive years, the inflammatory obstruction of the vagina becomes conflated with the allegedly unnecessary maintenance of robust labia, and women like Judy are allowed to progress to a point where "leaving it closed" is presented to them as a reasonable option.[14] In contrast, the physicians at the vulva clinic encourage patients to be proactive in maintaining their vaginal patency, or what they call "capacity." This can be done either through regular vaginal intercourse with a partner or, preferably, with the regular (daily) use of a therapeutic dilator (see figure 5). Dr. Robichaud typically prescribes two fifteen-minute sessions per day during which the patient keeps the dilator inserted.[15] While consensual and desired intercourse is certainly encouraged, the dilator is preferred because it can be used more predictably, with greater patient control, and with far fewer problems during the sometimes acutely uncomfortable flares of LP.

This treatment plan, in contrast to the one to which Judy had become accustomed, was derived from an investment in the anatomical and physiological well-being of the vulva and vagina—outside of any so-called need for vaginal penetration or sexual activity. Liberal prescriptions and applications of medications, close monitoring for undesirable side effects, careful instruction about the nature of LP, and treatment strategies geared toward maintaining as much vulvar and vaginal anatomy as possible were the material contours through which a patient at the vulva clinic came to experience her disease condition. These material strategies were obliged to a female genital imaginary in which optimal vaginal patency and vulvar contour are not options to be considered but rather anatomical ground to be preserved.

Judy's surgery was successful, and the next time I saw her she was bearing a mountainous basket of blueberry muffins for Dr. Robichaud, thanking her for the genitals that the clinic "gave back" to her. Of her (sexual) relationship with her husband, she told us, "We're in a great spot; the best in thirty years." Like Frances (see chapter 3), Judy left the clinic with a vulva in which she was now invested—a bodily imaginary through which she could better care for herself and with which she could generate an expanding number of genital possibilities, not least of which was a future of sexual intercourse with her husband. Her previous casual disregard for her genitalia, cultivated by at least two physicians and through an actively disinvested cultural milieu, had been replaced—at least for the time being—with the practice of getting up "pretty flippin' early" for the dilator sessions that she knew would help to preserve her genital vitality.

MANIFESTATION 2. A PUBLIC DEMONSTRATION, USUALLY OVER A POLITICAL ISSUE.

Although their tools, in the form of immune system modulators, were virtually identical, Judy's physicians wielded them with distinctly informed agendas regarding the use-value of female genitalia. For example, since the treatment of LP with modulators was still off-label at the time of my fieldwork, a vaginal applicator for their use did not yet exist. In fact, non-intravenous delivery systems of any kind were still relatively new, and the cream formulation most readily available was a rectal suppository. Apart from being anatomically contiguous, there are sharp differences between these two modes of administration: rectal suppositories help to distribute a drug systemically while sparing unpleasant gastrointestinal side effects, whereas vaginal preparations are often tailored for local delivery. Because immune modulators in the treatment of LP were meant to target both local tissue *and* general immune function, however, there was no reason that rectal preparations couldn't be used intravaginally, as long as the patient's affected skin and mucous membranes could tolerate the chemical base in which the drug was mixed.[16]

Unfortunately, a tolerance for pharmaceutical base creams did not offer smooth sailing for these patients, and more than one, for whom the drug was otherwise effective, recounted difficulties with its use. Anharrad, for example, was uncomfortable touching and manipulating her genitals and was reluctant to ask her husband for help. Her search for a vaginal suppository applicator led to a spate of embarrassing conversations with a spectrum of strangers—from pharmacists to natural foods store employees—during which she attempted to describe the "equipment" she needed in order to follow her doctor's orders. She eventually found something adequate at the natural foods market, but her frustration contributed to an ongoing conversation with Dr. Robichaud about not only the need for a vaginal applicator but the larger issues informing her inability to use her (or her husband's) fingers more directly. In one of these exchanges, Anharrad told us, "This is just another thing where women have to struggle to find something that works. I told the pharmacist that if I needed Viagra, you could sell it to me by the boxload."[17]

Complaints from women who *were* comfortable with vaginal insertion were of an entirely different order. Their stories revealed a misogynistic appropriation of their genitalia resembling what many feminists, in the wake of Rebecca Solnit's (2008) pointed essay "Men Explain Things to Me," have dubbed "mansplaining." Several patients told of having grown accustomed to the clinically unnecessary and suspicious questions regarding how their prescriptions were written. Often in front of other customers, pharmacists routinely (and sometimes with hostility) told these women that their doctors were mistaken, in that their medication was intended for rectal rather than vaginal use. Awkward conversations ensued, during which the woman would explain her vaginal condition; that it did, in fact, warrant the use of *this* drug in *this* way; and that, while she appreciated his concern, there was no need to alter the prescription.[18]

On the surface, these pharmacists were simply doing their jobs: ensuring that the drugs they dispensed would be used by their patients in the safest and most appropriate manner. This occasionally involved questioning the prescription itself, including the dosage, drug, amount, or delivery system, a scenario made more likely with an off-label prescription. Indeed, given that up to 20 percent of pharmaceuticals are prescribed off-label (Sanghavi 2009), even the most inquisitive pharmacist can likely not keep abreast of them all and, in this regard, I hold pharmacists only somewhat more accountable for this particular ignorance than I do a well-educated friend or colleague. But in the case of LP, (male) pharmacists' lack of information about both the condition and its treatment blurs the lines between inappropriate and ill-informed questions. And I suggest that these unstable boundaries—between careful dispensation, medical misogyny, prescriptive practices, and drug delivery systems—index the pernicious breed of vulvar *in*consequence under investigation in this book. The wavering line between a lack of access to accurate information (e.g., about intravaginal use of tacrolimus) and an active disinvestment in the genital integrity of women like Judy is made acutely manifest by these patients' experiences and reflects the deep-tissue discourses to which these apparently superficial practices are obliged.

Hiding in Plain Sight

> JOAN: She says, "How long [ha]s your skin look[ed] like this? . . . It's white." I said, "I don't know. I don't know. . . . I don't go down there and look."
>
> —From my field notes

Though more surgically dramatic, Judy's loss of labial contour was not measurably worse than was Mary Hudson's, the patient whose LS I have previously described. Like Judy, portions of Mary's labia and clitoral hood had effectively disappeared without her full comprehension, but because LS is significantly less inflammatory than LP, Mary remained unaware of these changes until a nurse practitioner noticed them during a routine annual exam. Joan, however, who recounted seeing a provider who was equally unsettled by her eroding vulva, arrived at the VHC with more missing than her labial contour. As I demonstrate in this section, Joan's story reveals that vulvar disinvestment also intersects with age and socioeconomic class.

Joan arrived at the clinic for what she called a "vaginal screening," telling Dr. Robichaud, "My doctor wants me checked for cancer. [That's] all they'll pay for." Joan had indeed been approved for only one visit to the clinic by the state health insurance plan, and the paperwork she'd been given clearly indicated that a pap smear and *vaginal* biopsy were all that would be covered. Based on both her treatment history and the description of her symptoms ("This itching is driving me crazy!"), Dr. Robichaud wanted to perform a vulvar biopsy, as she suspected lichen sclerosus. Unsure that the hospital would be reimbursed, however, she and the clinic manager agreed that it would be best for Joan to return with the correct approval rather than for the clinic to try and obtain retroactive reimbursement.

Fortunately, Joan was able to do so, and two weeks later, I observed her vulvar biopsy. Dr. Robichaud had prescribed a mid-potency steroid at the previous visit, and Joan reported that her itching was now only bothering her "from time to time." As Dr. Robichaud began preparing for the biopsy, she and Joan had the following exchange.

> DR. ROBICHAUD: These are your labia majora, these thicker ones with hair. Your labia minora are gone. And your clitoris—I can't see [it] anymore.
> JOAN: It's probably gone too.
> DR. ROBICHAUD: No! It's not gone. It's hidden. [She

explains that the scarring that has happened is because of the LS-related inflammation.]

JOAN: Can anything be done about this?

Joan's first two visits to the clinic demonstrate the limits of even the most invested provider. Dr. Robichaud's decision to forgo Joan's biopsy at her first visit was obliged to insurance and health care industry practices that have the authority to shape the definition of good health. In this case, it was cancer-free—but not completely contoured—genitalia that garnered the financial support of the state insurance plan. Joan was left to conclude that the preservation of her labia—*in and for themselves*—was a luxury that the state could not afford. And although Dr. Robichaud exhorted Joan to maintain an investment in the material reality of her clitoris ("No! It's not gone."), her decision to send Joan home could have easily sent Joan's vulva into even deeper hiding.

When I interviewed Joan several weeks later at her trailer home in a rural part of the state, she told me about a relationship history in which sex had always figured prominently and pleasurably. Indeed, Joan had not one but several sets of nude photos of herself that had been taken by previous partners, and she told me about them with far more of a knowing smile than any sense of embarrassment. Rather than using this fact to set up a contradiction or disingenuousness to Joan's claim that she preferred to steer clear of her vulva, I want to suggest instead that Joan's lifetime of bodily pleasure is perhaps what enabled her to persist in securing the coverage for her vulvar biopsy and to return for the care that she knew was available at the VHC. At just her second visit, and in connection with Dr. Robichaud's attentive counseling, she poignantly asked whether "anything [could be] done about this." Unsurprisingly, however, Dr. Robichaud's reply—and ultimate investment—was compromised by the economic context of the situation:

DR. ROBICHAUD: [If LS is left untreated] the labia can stick together. And then we have to do surgery to separate them. If you treat it, then we don't have to do this.

JOAN: Well, if my insurance will pay for it. If not, I'll just have to live with it.

DR. ROBICHAUD: Well, I would hope that your insurance will pay for some of [it].

It is easy for most of us to comprehend that the state health insurance plan is not caring for Joan (or her vulva) adequately. But locating and *holding accountable* the deep-tissue discourses through which that "care" is conceived and administered is a more formidable task. Joan's having returned to the vulva clinic and undergone the appropriate biopsy for her symptoms did not give her back the pieces of her vulva that were already gone. It is also likely that, without the kind of clinical and emotional support that allows women to effectively resist these pernicious discourses (Connor, Brix, and Trudeau-Hern 2013), Joan will lose more of her labia in the years to come. Finding the connections between this absented flesh and the all-too-present cultural assumptions about genitalia for which women can learn to "make room" (Brown 1995, 101) allows us to more concretely imagine how the material *presence* of nondiseased female genitalia is nonetheless fundamentally informed—and perhaps experienced—as a profound and censored *absence*.

The vulvas with which Judy, Joan, and Mary Hudson struggled to live are bodily instantiations of overlapping discourses regarding female sexuality, excess, reproduction, heterosexuality, "health" (Metzl and Kirkland 2010), and genital normativity. Paternalizing pharmacists, conservative dermatologists, and narrow-minded insurance administrators reveal particular slices of the vacuum-like silence through which many patients live their symptoms, but they are fragments readily perceived by a critical and ethnographic engagement with these disease conditions. This book details the disavowing and *active* nature of discourses that rob many women of a genital "capacity," indexed here by the vaginal patency from which Judy's first physician encouraged her to disinvest. Delineating the social spaces through which this invisibility is made manifest helps us to see that the silence in which vulvar pain is experienced might be more accurately described as a censored story. Without explicitly proclaiming that they "don't give a shit about" the genitals of these women, institutionally located actors convey this sentiment in their everyday acts of evasion, erasure, and disparagement. The vulva's absence from the cultural landscape indexes incompetence at the structural, rather than individual, level (Metzl and Hansen 2014), making it all the more challenging to locate and hold accountable the spaces through which this invisibility is rendered. Seeping into the discourses and practices of even those officially charged with the task of bodily care—notably, physicians and women themselves—vulvar disinvestment hides from sight as readily as the diseased genitals of my informants.

A MEDICAL CONFERENCE

MANIFESTATION 3. THE STATE OR CONDITION OF BEING SHOWN OR PERCEPTIBLE.

On the morning of October 27, 2004, I walked up to a small table at the Hyatt Regency in Atlanta, Georgia, and, in exchange for my name, was given a binder stuffed with an agenda, supplementary articles, faculty biographies, and plenty of blank space for taking notes. I had just officially checked in to *Vulvodynia and Sexual Pain Disorders in Women*, a conference I had learned about from the NVA's online newsletter and that promised to be both state of the art and the first of its kind.

Technically, some of the researchers present that day had convened once before. Catalyzed by the NVA's persistence, several key vulvar experts had met with both the NVA and representatives from the National Institutes of Health (NIH) in the fall of 2003. At that time, the NVA's goal was to secure a legislative call for vulvodynia research funding, and the NIH solicited the expertise of vulvar specialists in order to accurately gauge their interest in sponsoring such a bill. That meeting, however, which figured significantly in a 2007 House Appropriations bill that directed the NIH to develop a National Vulvodynia Awareness Campaign (NVA 2013), had not been open to the public, meaning that the conversations that transpired and the conclusions that were reached were not available to the bulk of clinicians who were treating patients. Within the field of vulvar pain, then, the conference I had just checked in to marked an important beginning. An expanded group of experts was available to discuss vulvar disease from innumerable angles: bench research; sexuality; demographic and epidemiological statistics; psychotherapy, biofeedback, and cognitive-behavioral therapy; neurological, immunological, dermatological, and pain-related physiology; concomitant urological conditions; and, to a lesser degree, treatment options. Moreover, and perhaps most importantly, there was an audience full of clinicians—in various states of frustration, confusion, and optimism—ready to learn from them.

To my knowledge, I was the only social scientist in the room that day, and I quickly began writing field notes, the first of which noted the "overwhelming number of women" in the audience. Many, if not most, of the men and women sitting at the front of the room, as well as those described in my program guide, were researchers whose names I had come to know while I learned about vulvar pain. Among these icons was a physician named David Foster, to whom I paid particular attention. Foster had become important to me because of a 2002 article he published in *Obstetrics and Gynecology*, the goal of which was to provide women's health clinicians with a comprehensive overview of vulvar diseases.

Around the time that Foster's article was published, I was developing an argument about the relationship between what I now call genital alienation and US women's increased risk for vulvar cancers. Since the initial presentation of vulvar cancer typically involves skin changes that can best be monitored by a woman familiar with her vulvar "baseline," the condition can worsen appreciably if not detected early (Duarte-Franco and Franco 2004; Labuski 2013). Gynecological clinicians interested in vulvar disease conditions understood that women were reluctant to perform genital self-exams, but they did not (in my mind) adequately grasp the cultural habitus in which this reluctance was stubbornly lodged. Thus, their exhortations, to both patients and other clinicians, for more vulvar self-exams (Lawhead 1990) fell on ears that were not so much deaf as ill-equipped to comply.

Opening with the observation "The vulva most clearly defines the female phenotype and yet, the female patient commonly knows less about her vulva, in health and disease, than any other part of her external anatomy" (2002, 145), Foster's article gestured toward a correlation between women's bodily ignorance and the likelihood that vulvar disease would be detected at an advanced stage; I had expected him, therefore, to direct at least a few of his thirty minutes on the morning of the conference to the affective dimensions of vulvar pain. My expectations were fortified by my experiences with the VHC clinicians, who nurtured my belief that any physician or researcher specializing in vulvar pain was at least partially guided by a sense of preventive urgency. But given the titles of Foster's two papers that morning, both of which were particularly physiological, I might have been better prepared for the direction of, and my eventual disappointment with, his presentation.[19] As I struggled to keep up with slides and data based in molecular biology and genetics, I noticed that I was battling a separate set of fieldwork-related concerns: that I was having trouble really *caring* about the importance of his findings. What, I began to wonder, did interleukin and capsaicin have to do with the lived experience of vulvar pain? (How) would this research make its way into my ethnographic account of vulvodynia and VVS?

In retrospect, my concerns were ill conceived, as I was—and remain—well aware of the role that clinical research plays in the experience of a disease condition, most specifically in the contributions it makes to diagnostic markers, screening tests, and treatment protocols. At the time, however, I found myself resisting Foster's paper, even cultivating the recalcitrant attitude that produced this field note:

> Dr. Foster (and these other MDs) are attempting to make vulvar pain better, but they are doing so through molecular structures that have no part in the patients' narratives. These women are describing pain that literally prevents penetration by their husbands ("it's like a wall in there"). They are describing soreness after genital contact that is so severe that they need to apply bags of frozen peas or cans of frozen lemonade to their

vulvas for hours afterward. They describe what they understand to be redness around the opening to their vaginas, and particular spots that neither they nor their husband's fingers can go near without them flinching (at the very least). They talk about how things have seemed to get worse over time and about how if he "can get past that one spot," or if they are on top, it can be okay. They describe how unbelievably sensitive their skin is—how "nothing" can touch it. They say that their husband "doesn't fit" into their vaginas. Their nerve endings and immune systems are not (yet) part of these narratives.

Again, in looking back, it is easy to see that my own myopic lens on vulvar pain likely mitigated my ability to appreciate the salience of Foster's research, not only for the clinicians in the audience but for the patients as well, patients on whose behalf his research was being carried out, about whom my field notes were allegedly concerned, and who were more than adequately represented by advocates who had worked to convene the conference and bring Foster to Atlanta.

Foster's work with capsaicin—the heat-producing chemical found in chili peppers—was a trendy topic in neuropathic pain research at the time, evidenced by the NIH funding he had already secured (Foster, Dworkin, and Wood 2005), as well as by the attention paid his presentation by the providers, patients, and advocates in the audience. But as his presentation continued, I could not shake the sense that something was amiss: Would any of his research or treatment protocols correct the genital ignorance he had so straightforwardly lamented in 2002? Had I imposed the sense of dismay through which I'd always interpreted his article? Did he understand vulvar ignorance as a condition worthy of (his) clinical intervention, or was pain the only problem he perceived? In other words, in focusing on the genetic, immunological, and neurophysiological aspects of VVS, was he (and were other researchers) losing sight of the vulva *itself*?

In *The Birth of the Clinic*, Foucault contends that "in order to know the truth of the pathological fact, the doctor must abstract the patient" (1973, 8), and that

> the doctor's gaze is directed initially not towards that concrete body, that visible whole . . . that faces him—the patient—but towards intervals in nature, lacunae . . . [and] distances in which there appear . . . "the signs that differentiate one disease from another, the true from the false . . . the malign from the benign." (8)[20]

As I listened to the papers delivered that day, by the most important researchers and clinicians working with vulvar pain at the time, I became increasingly aware that our ideas about vulvar pain's "true" aspects were differently informed. The malignant forces that *I* see operating on and in afflicted bodies do not exclude molecular and physiological occurrences, but they are simultaneously and fundamentally *obliged to* the states of alienation and inconsequence that I describe in this book and that are culturally produced and sustained.

"Medical looking is not naïve," argues Katharine Young (1997, 123), but is, rather, a product of the epistemic shift described by Foucault (1973) and characterized by a change in perception. In the "modern" body, "signs, which used to be clues to the past . . . the present . . . or the future . . . , became symptoms, manifestations, localizations, *instances of the disease* (Young 1997, 123; my emphasis). The tension of the conference, at least for me, was that in localizing vulvodynia and *VVS* to nerve endings and measurable pain responses, the "clues" to the pasts, presents, and futures of *any* woman with a (disparaged) vulva were being excised from the broader perception of vulvar pain. The gaze of these clinician-researchers was indeed far from naïve, structured by both epistemological trends as well as by the proliferating funding opportunities on display in the Hyatt Regency that day. Presenting researchers had been handpicked by the NVA and relevant NIH personnel, all of whom had active investments in moving vulvar pain research in particular directions, notably *away* from the marginalized realm of the psychological-affective. In emptying lived experience from the conversations that day, the conference reduced vulvar pain conditions to a list of "ingredients" (Michael and Rosengarten 2012, 11; see also Nielson 2012) that could be subtracted from or added to diagnostic schemas, often without meaningful input from patients who suffer the symptoms.

As the day progressed, and I sat through lectures about the plasticity of the nervous system, immunogenetic analysis, vulvoalgesiometry, cytokines, and steroid hormone mechanisms, I became aware that I was witnessing—in real time—the paradigm shift that I have previously described: from an invisible and psychosomatic embodied trauma to a pain condition whose physiological location and etiology could be mapped with increasing precision. As one researcher presented the first official "algorithm of care," followed by an expert (and rather spirited) panel discussion about nomenclature and diagnostic categories, I noted that I was also witnessing the medical *realization* of vulvar disease. The algorithm, we were told, was about to be published (Haefner et al. 2005) and would significantly affect subsequent diagnostic and treatment protocols; we in the audience had been treated to a sneak preview of what would now be the standard of care against which anyone treating vulvar pain should compare their own efforts. It cannot be overstated that such a standard—essentially defining the state-of-the-art, or gold standard, of care—*is* medical reality. In essence, vulvar disease was made manifest in the ballroom of the Hyatt Regency that day, between the hours of 7:30 a.m. and 4:30 p.m., to be exact. A new brand of vulvar pain had officially emerged, based in significant part on the relations and practices through which it was enacted (Mol 2002; Michael and Rosengarten 2012). This new clinical "reality" would be maintained by the perfusion and circulation of these standards and discourses among the general population of gynecologists.[21]

A New Vulva?

Anthropologists and others have long demonstrated that many contemporary "life" scientists understand the bodies with which they work in intriguingly plastic terms (Duster 2006; Goodman, Heath, and Lindee 2003; Lock et al. 2007; Rabinow 1997): a postmodern bodily milieu "where the artifactual and the natural have imploded" (Haraway 1997, 245) and where bodies redefined by nucleotide sequences and neurotransmitters are open to novel interpretive frameworks that can reduce and reify but also *exceed* our received cultural categories (Gilroy 2000; Reardon 2012). From this perspective, it is productive to examine my experience at the conference through an alternative gaze.

Surrendering to—or evolving with—these contemporary scientific narratives, I am cautiously enchanted with the potentially liberating nature of technoscientific narratives and with the ways that their basis in the so-called natural science of biology seem to cleverly cloak projects and discourses that are decidedly artificial. Haraway (1997) reminds us that genome projects, for example, "produce entities of a different ontological kind than flesh-and-blood organisms, 'natural races,' or any other sort of 'normal' organic being" (247). Read one way, this narrative can free up the essential nature of *any* kind of body—black or white, female or male, diseased or "well." That is, bodies manipulated and interpreted as "bits of life" (Smelik and Lykke 2008) can be(come) available for a multiplicity of readings, perceptions, and interpellation; per Haraway (1997) and Barad (2007), we might locate these scientific practices in a diffractive field, allowing them to shuffle and reconfigure the most basic elements of the bodies in which we currently live. Where, then, would a body start? What defines its pleasures and its pains? Its cohesion and disintegration?

In allowing that these modes of inquiry might facilitate ontologically distinct apprehensions of the body, and in making room for Dr. Foster's having "abandoned" the flesh of his patients' vulvas for a materiality of a different kind, I find myself with a new set of questions. Regarding the connections between the modern and postmodern bodies theorized by science studies scholars and the apprehensive shift that I witnessed at the conference: Does the "state of the art" line of inquiry evident at the conference mean that nerve pathways and immunogenetic markers constitute "the *new* vulva"? And if so, (how) can I make this vulva meaningful? Can I come to *care* about it?

My reluctance to answer this question in the affirmative stems from an acute and experiential knowledge of how the "bodies" in question are located within a specific cultural context, in this case, one of routine contamination and disavowal. The apprehensive shift that I witnessed in Atlanta—the same shift evident in medical journals, funding requests, and patient-physician dialogues—is beginning to make *absent* the site through which healing and reconciliation can occur. If, as I believe, vulvar pain does not emerge outside the cultural experiences of inconsequence and alienation, then treatment strategies (and the perceptual pathways through which they are developed) that bypass the anatomical vulva do not offer patients an effective path to recovery. Indeed, they actively participate in the practices of erasure that compound both the silence and the suffering of symptomatic women.

Freud argues that a "conscious presentation" requires the coexistence of "the thing plus the . . . word belonging to it" while "the unconscious presentation is the presentation of the thing alone" (in Grosz 1994, 29). My vulvar dilemma articulates with and remains unaddressed by this set of propositions in that what was presented at the conference seemed to be the word without the thing. In a room full of vulvar experts, there was no linguistic reticence. And yet I believe that I witnessed a trade-off, one in which the thing itself was exchanged for the words that belong to it. I am unsettled by this displacement, and the pages of the conference notebook do not soothe me. Full of PowerPoint slides, bulleted lists, and faculty biographies, I find only one vulva amid a proliferation of images—graphs, text, schematics of pain pathways and neuroimmunological mechanisms, MRI scans, algorithms, symptom scales, and photographs of diagnostic instruments. Indeed, Dr. Foster's discussion of capsaicin, which involves an injection of this chemical into the forearms and feet of symptomatic women, includes not one but three photographs of *feet*. Has the vulva been erased by a state-of-the-art convocation designed to make it visible? And if so, how will the genitals of symptomatic women be regarded by those whose gold standard has now been constructed through its discursive *dys*-appearance (Leder 1990)?

My disgruntled conclusions about the perceptions through which these emerging standards were developed were based on one more important factor: the complete absence of physical therapists from the conference faculty and invited speakers. Still green in the field, I might neither have noticed nor questioned their absence if the therapists in the audience had not made their presence known through their numerous questions and comments. Indeed, it is fair to say that almost half of the audience interaction that day was with physical therapists, most of whom had tough, concrete, and experience-based questions for the experts. One woman in particular, who posed more than one of these queries, shared her knowledge with other audience members when one of the panel members was stumped.

Although I had no direct experience with PT at the time, I had nonetheless begun to perceive that the physicians at the vulva clinic did not fully comprehend the details of how and why PT was so effective, beyond its capacity to massage out pelvic floor myalgia. As the epistemological gaps between questioner and expert became more obvious and even awkward at times, I wondered about how, when, and why physical therapists had been relegated to the sidelines. And though I take up these questions more fully in the following chapter, I sketch out a brief set of answers as I conclude the present one.

The marginalizing of physical therapists may reflect no more than distinct levels of vulvar "distance" maintained by different clinical orientations toward disease and the body. But the second-class status of PTs also reproduces a broader set of divisions between the mind and the body, and between classical and grotesque bodies (Russo 1995). Researchers deploy masculinized and scientific *minds* to explain vulvar pain's complex and neuropathic processes, tolerating (as do Wilson's gastroenterologists) a complementary role for the flesh-and-blood—and emotionally contoured—*bodies* that PTs literally take into their hands. In the space of the clinical conference, the coarse and manual details of how PT "works" remain epistemologically subordinated to the abstracted scientific etiologies being "discovered" by researchers like Foster.

Some authors (Bernheim and Kahane 1985; Masson 2003) have argued that Freud developed the unconscious at the expense of the sexually abusive stories of his hysterical female patients. Rather than believe that such a large number of women could have been mistreated at the hands of husbands, fathers, or family friends, the argument goes, Freud chose to make the narratives of abuse symbolic rather than real, effectively displacing the cause of neurotic hysteria from structural patriarchy and idiopathic pedophilia to the individual and psychically diseased minds of (victimized?) women. I find it hard not to marvel at the resonance that such a revisionist history has with the social and clinical history of vulvar pain. In the second half of this century, the perception of symptomatic women has swung from that of "probably abused" victim to a "not necessarily abused" conundrum, to a hyperalgesic body for whom a history of sexual abuse is increasingly irrelevant. In reviewing this history, I cannot help but wonder whether the clinical realization of vulvar pain is yet another case of an authoritative and masculinist medicine that can listen to but not *hear* the stories that women are trying to tell.

This line of inquiry is further complicated by the conflicting desires and refusals of symptomatic women, many of whom have insisted that there is no room for an abusive past in the etiological recounting of their pain. But, as I showed earlier in this chapter, an expanded definition of "abuse" has led to a proliferation, rather than a reduction, in the number of women who can be diagnosed with vulvar pain (Harlow and Stewart 2005). If vulvar pain conditions are even partly the sociosomatic eruptions that I propose—in other words, culturally contoured symptoms that, long silent, have now *"forced their way into consciousness"* (Freud 1917a, 345)—then I believe this is a lead worth following. We must continue, in other words, to look for ways to infuse the definitions of vulvodynia and VVS with the embodied pasts, presents, and futures of symptomatic women, many of which contain significant amounts of unwanted genital experience, vulvar alienation, and bodily disintegration. To that end, I use the next chapter to analyze the relative ability of symptomatic women to integrate corporeally contaminated pasts with clinical presents from which such histories are increasingly detached.

5

INTEGRATION

Coming Together or Falling Apart

―•―

GETTING BETTER BY PUSHING THE LIMITS

The three women who "got better"[1] during my time in the field—Daphne, Libby, and Jessica—worked with the three physical therapists who shared a practice and to whom the VHC clinic turned most frequently: Cathy, Hanna, and Joy. Cathy, who owned the practice, lived a busy but balanced life that included five children, a working husband, regular bodywork, and a family band that competed twice a year. She admitted to a past that was significantly more hectic—jogging during her lunch hour and working six long days a week in order to build her practice, all while actively raising her children—and she used this history to commiserate with patients who were struggling to find time to do the homework she routinely assigned. Cathy recruited and hired like-minded therapists to work in her office, women who understood the importance of making regular and sustained connections between daily life, emotional and physical pasts, unforeseeable circumstance, and the bodies that registered the impacts of these events. In talking with her once, I commented on how remarkably effective she was, and I wondered aloud about how she both elicited and managed the very poignant narratives that accompanied vulvar pain. Cathy told me that she could not imagine caring for her patients without addressing these facets of their symptoms, and then contextualized her approach by adding, "Oh, [but] we're definitely pushing the limits here."

Cathy's assertion goes directly to the heart of why these therapists were so successful. In order to make the kinds of differences that mattered to their patients, they needed, like the vulva itself, to exceed professional and culturally circumscribed boundaries. In pushing the limits of PT, they offered their patients a view to something new: a body that was deeply integrated and comfortable in its skin, muscles, breath, and idiosyncratic rhythms. When Cathy's interventions brought a patient to a place of greater physical tolerance (e.g., of Cathy's finger inside the patient's vagina), or when she was adjusting pillows around a back or pelvis, she would ask, "So, how does that feel?" When the patient replied, "It's fine," or that she was "okay," Cathy responded with a question meant to move her patients' attention beyond the task at hand. "Yeah, it's okay," she would reply, "but how can it be *better*? Sure I'm comfortable, but how can I be *better*?"

How, indeed, can female sexuality be better? If the lived experience of vulvar pain indexes the fragile and disparaged state of women's genitalia in the contemporary United States, then (how) can "getting better" provide for their meaningful incorporation? And what else must be addressed in order to live in such a body? To answer these questions, we must explore the social and material spaces through which the vulva's inconsequence is routinely realized. This chapter brings these spaces together in the forms of patient PT narratives, clinical protocols, neurophysiology, medical research, cultural discourse, and feminist theory. These stories and analyses reveal that the body images (Grosz 1994; Schilder 1950) of many women—epitomized by those with vulvar pain—do not contain their complete genital anatomy and that "getting better" requires that fragmented female genitals come to matter in order to be recuperated.

Pelvic floor PT is a deeply personal experience, and patients, particularly in Cathy's practice, were encouraged to talk openly about the emotional aspects of their pain. As symptoms improved through regular sessions and homework, therapists began to engage each woman in frank dialogue about sexual behavior, helping to cultivate strategies that could maximize her chances for successful penetrative sex.[2] PT involves the regular penetration of a patient's vagina, either with her own or the therapist's fingers, a biofeedback sensor, and a therapeutic dilator; it can also include rectal work if a woman's "holding pattern" involves this related musculature. This work was arduous, and each time I was invited to observe a session, my understanding of what else these women faced in their confrontations with vulvar pain deepened considerably. I use this chapter to explore the content and context of this "what else." That is, I use PT and concepts from corporeal phenomenology to discuss the numerous facets of a woman's life that can either come together, fall apart, or remain unexamined during her attempts to alleviate her genital and sexual pain.

VHC physicians divided vulvar pain into three main categories: *vulvar vestibulitis syndrome* (*VVS*), *vulvodynia*, or one of three *lichens*. The latter two conditions typically arise later in a woman's life, up to and including her postmenopausal years, meaning that these patients—although their pain may now be debilitating—have experiential knowledge of previously pleasurable genital and sexual contact. Women with VVS, on the other hand, usually have no such corporeal reference point and have, at most, memories of a few fleeting weeks at the beginning of their marriage or sexual relationship when sex didn't hurt. For the typically older women with the latter two conditions, the reality and longevity of a past without sexual pain—and the desire to have it back again—enhances their ability to seek attention quickly, which often secures an earlier and more accurate diagnosis. Without such nostalgic biofeedback, patients with VVS struggle, often for years, with fitting the experience of searing pain into popular and allegedly friendly advice that (a) penetrative sex normally involves some degree of pain; (b) she just needs to "relax"; or (c) there is no help to be had. Unfortunately, and most pertinent to the subject of PT, by neglecting the injured skin of her vulva, each of these patients sets up the conditions through which her vagina and pelvic floor will also begin to burn and reject penetrative contact (De Andres et al. 2015).

Physical therapists perceive a proportional relationship between the extent of pelvic floor injury and the length of time a woman's body has been in a compensatory mode, where the *pubococcygeal* and *levator ani* muscles have been conspiring to protect the body from behaviors associated with insult or injury. Like any muscle grown tight from prolonged contracture, the hammock-shaped pelvic floor must be loosened, stretched, and retrained in order to reverse accumulated damage. Depending on how long she has gone without a diagnosis or how vehemently her muscles have reacted to the presence of pain, a woman with vulvodynia or one of the lichens may need some of this work in order to supplement a clinic-based treatment plan that targets her skin. But it is primarily the women with VVS, whose genital and sexual desires have been hijacked by the ever-present possibility of contact-related pain, who most need what physical therapists have to offer. Indeed, the coming together of the right therapist and an appropriately ready patient can defer, and occasionally obviate, the need for skin-based regimes such as surgery, laser ablation, and topical anesthesia. As I show in this chapter, however, accessing this kind of care can be stymied by a host of variables, including the vagaries of clinical medicine and questionably sympathetic partners.

In the Clinic: The Stratification of Better

In referring symptomatic women to PT, VHC clinicians began a dialogue through which they hoped to enlist patients in their own recovery. When they collected a patient's symptom history, the doctors actively disavowed the psychosomatic discourses through which so many women had come to understand their pain: "I believe you," was Dr. Robichaud's patterned and first-line response to her new patients, women whose bodies had betrayed their sexual expectations to an extent that they themselves could barely believe. "Your pain is real," she continued. "It is *not* in your head."

After using the Q-tip test to delimit the extent and intensity of a patient's skin pain, the doctors performed an internal exam to evaluate pelvic floor involvement. And though their sympathy and respect for each patient's subjective experience of pain remained central to these interactions, it was during this part of the exam that clinic doctors began priming their patients to take responsibility for their (future) pain. By validating the reality of their patients' pain and simultaneously explaining how prolonged muscular contraction could lead to a burning sensation confusingly similar to vestibular pain, Drs. Erlich and Robichaud established a physically demonstrable connection between a portion of a patient's pain condition and her own conscious action.

Any patient whose assessment suggested muscular involvement was given a formal—and therefore billable—diagnosis of pelvic floor myalgia (muscle pain), in addition to her vulvar diagnosis. Aside from making pelvic floor pain a clinical reality, this billable code legitimated the need for PT to insurance companies.[3] After consulting a list of local and statewide therapists, VHC doctors typically recommended two or three to each patient, a medical decision that was more art than science. Drs. Robichaud and Erlich used their quickly gleaned impressions of each patient to predict which PT approach and personality would best meet her needs. Their first criterion was typically geographic, as many women had traveled either across the state or from another state for their consultation.[4] Some areas of the large region from which patients were drawn were not served by anyone on the list, but more than half of the patients I met were given the contact information for at least one therapist within seventy-five miles of their home. Since the clinic was in the largest city in its state, patients from the greater metropolitan and suburban areas were often able to choose from several options.

But what appears to be a simple case of administrative logistics was also an index of how messy Vulvar Disease can be, often undermining the clinic's best efforts to establish a uniform standard of care. With no control over the unequal access that symptomatic women had to adequate clinical care, the VHC physicians attempted to sustain the Women's Health Center's "commitment to excellence" by fashioning treatment plans that could extend *their* expertise across the state or region. The awkward reality that had to be faced, however, was the dearth of professionals that could sufficiently manage these notoriously confounding symptoms, a fact that meddled with physicians' attempts to secure a treatment they believed to be integral to their patients' recovery. Although none of the physicians or therapists ever explicitly acknowledged this to me, what was uncomfortably obvious during these postexam negotiations was that the patients—and their chances for recovery—existed on an extremely slippery and stratified playing field, with a sizable number unlikely to be, in Cathy's words, "better" for their efforts. Indeed, some may have felt worse, having glimpsed how arbitrary configurations of insurance, geography, and luck-of-the-draw affected the resolution of vulvar pain.

At Home: Here and There

Some details about thirty-two-year-old Mya, who had been in pain for about nine years when I met her in December 2004, both illustrate and deepen these issues of access and disparity. By revealing the compromised genital integrity that Mya brought to the exam table, I want to broaden our notions of health care access. That is, regardless of what is diagnosed, prescribed, geographically available, and covered by insurance plans, if a woman cannot *access* the parts of her body that require treatment, she is unlikely to understand or pursue the recommendations of her physician. When I interviewed Mya in February 2005, she had not made arrangements for PT, nor did she plan to. She also did not have another appointment in the clinic and told me, "I don't know what I'm supposed to do next." This confusion resulted from a profound underintegration of Mya's clinical needs, her available resources, and her affective access to her genital body. Without any one factor tipping the scales, a pattern of poorly understood—and only partially developed—treatment possibilities circulated and fragmented around Mya's pain.

Mya told Dr. Erlich that she had been symptomatic for two years, and that her vulvar condition had been evaluated (as such) for about that long. But she told me in our interview that her pain began when she was twenty-two, ten years before her appointment at the clinic. Mya did not have VVS, and she had been able to participate in, and even enjoy, sexual intercourse for a number of years before the onset of her disease. Mya's pain was more typical of vulvodynia, in that she described pain both with and outside of contact and penetration. "It can happen when I'm doing the dishes.... Sometimes it's as if someone is stabbing me," she told Dr. Erlich and me at the beginning of her initial visit, two days before Christmas. Mya also provided a history of deeper *dyspareunia* (pain with thrusting), but her description of this was vague; in terms of sexual disruption, she seemed to prioritize the symptoms that she described as "outside." Unfortunately, "because of the way [she] was brought up," Mya struggled to communicate these symptoms to her physicians—the main reason that her symptoms were not clinically evaluated for over seven years.

For Mya, "at home" partially signified a mother that was "like a fifties mom; everything had to be so perfect," a mother who told her as an adolescent that if she used menstrual tampons she would no longer be a virgin (Carpenter 2005). Mya attributed her inability to accurately describe her pain to this familial history, repeatedly telling me that it was "how I grew up," but I believe this assessment is only part of the story. Individualistic in focus, her narrative casts blame solely inward, as she takes personal and familial responsibility for the third of her life that her genitals have been in unmitigated pain. Significantly for Mya, "at home" was also the place where she lived with her partner, a man who did not believe that her symptoms were real and who called her a "lunatic" on more than one occasion in relation to her complaints. Although my interest here is not in culpability, I nevertheless want to hold accountable the numerous cultural locations that both enable and sustain this kind of self-, or local, blame, as well as the particular brand of silence in which Mya is engaged. These are the sites where vulvar dis-ease takes virulent hold, making "at home" far less safe and supportive (Harlow and Stewart 2005) than our prescriptive advice needs it to be (Mulla 2014).

Having successfully transitioned from health care provider to anthropologist, it is easy for me to now see (and accept) that even the most sensitive, feminist, and activist health care is just one discursive fragment through which a woman comes to know her body. As a clinician, I was plagued by, but did not have time to address, the questions that now fuel my academic research, questions about the other sources of a woman's bodily knowledge and the sites where she encounters them: What happens, for example, when a woman goes "home" from a successful clinic visit, and (how) can an anthropologist intervene into processes that might undercut the education and empowerment she might have gathered there? During my first years as a clinician, I understood "home" as an amorphous entity that included—in some loose order—the kinds of relationships a woman was in, her family traditions and demands, and, perhaps, her desensitized exposure to popular media. Immersion in anthropology and critical gender studies has sharpened and formalized—though not fundamentally changed—this definition of home: I still want to know what happens *there* when the clinic is *here*.

In the *here* of the VHC, the physicians strove to care for their patients' bodies in the context of their larger lives: they made those lives present and they promised, in Dr. Robichaud's words, "not to abandon" their patients in the time and space between their visits. Both physicians also employed the phrase "at home" as code for sexual intercourse. When investigating the nature and extent of a patient's pain, they often instructed her to point to the areas that had been most sensitive, asking her to show them "where it hurts *at home*." Drs. Erlich and Robichaud also used liquid lidocaine—applying it in healthy layers to the vulvas and vestibules of symptomatic women—to mitigate the pain of their speculum exams. In doing so, they called attention to its anesthetic properties and wrote each woman a prescription for a supply to be used "at home" (i.e., during attempts at coital penetration). This practice in itself is not problematic; clinicians employ a variety of euphemisms in order to avoid offending their patients and, at times, are just being verbally creative. But for Mya and her fellow patients, "at home" takes on a new set of meanings—the kind that I wondered about during my days as a clinician and the kind that cannot be remedied with the application of a topical anesthetic. If "at home" signifies what happens *there*—outside of the clinic and its genital attention—then treatment modalities need to go both deeper and farther (Mulla 2014). They need to, in Cathy's words, "push the limits" of what medicine is expected to do if they are to alter women's access to recovery from genital pain.

Fortunately for Mya, there was a physical therapist in her area (a five-hour drive from the clinic) who was learning how to work with vulvar pain. But this lone practitioner proved no match for the cultural contexts of scarcity, erasure, and denial with which Mya's genitals were so thoroughly perfused. Dr. Erlich tried her best, ensuring that the therapist was accepting new patients before she sent Mya away with little more than a bottle of lidocaine. She personally located the pamphlet for the practice and, seeing the therapist's photo inside, asked Mya, "See? Doesn't she look like she could help you?" Mya was unenthusiastic—"not terribly charmed," according to my field notes—and said only, "What a thing to go to physical therapy for. . . . Is this the only way it's going to get better?"

As uncomfortable as it might be to admit, well-intentioned clinicians are particularly frustrated by patients who respond in this way, reluctant or unwilling to do what we believe to be in their best interests, particularly when they *claim* to want to get better. But requests for relief from vulvar pain are acutely enmeshed in so many years of shame, confusion, and bodily betrayal that they require a level of care that accepts and moves through this emotionally painful terrain. Mya's recalcitrance was far from anomalous, overdetermined by the factors under consideration in this book. Her ultimate refusal to participate in PT once she got *home* was a return to her familiar frames of isolation and genital alienation. She, like Clair, could not "even imagine" sharing her bodily experience with anyone but a physician or sexual partner; she also voiced concerns about the nature of a small town and the consequent fears of public exposure that such a relationship would risk. I will not speculate on whether, like Nikki, Mya could have benefited from physical therapy if she could have "just relaxed" about it; such a question—individual in focus—does not take into account the cultural factors that limit women like Mya's access to genital well-being. Rather, the bodily reticence that constrained Mya's recovery efforts was performed within a social habitus that effectively curtails matters of female sexuality, structuring what counts as *going too far*.

TOO MUCH OR "A LOT MORE THAN I AM"?

So, I ask myself, is this what my book is trying to do—confront this "totality" of vulvar pain in its cultural and historical context? My answer is that I think so. Is vulvar pain the embodiment of a cultural distaste for the vulva and for the excessive sexuality that its nonreproductive nature indexes? I'm a little less certain about this. Were Harlow and Stewart (2005) reluctant to fully embrace the results of their safety and support study because they exceeded the confines of medical discourse, because they were just *too much* to take on? Of that I am even less sure. Regardless of my ability to answer these questions, however, I pose them because they each address the themes of *integration* that orient this chapter: the fragmented and alienated spaces through which the vulva—in both material and discursive forms—is either accessed or ignored, including by women themselves.

"If it were my job to mathematically figure out which women despise more, being called a cunt or having one, I'd be hating life," writes Inga Muscio in her 1998 manifesto, *Cunt*. "I'm glad that is not my job" (27), she tersely concludes. Clearly, I am similarly vexed by the apparent complicity of US women in the erasures of their own genital bodies. Muscio imagines, and calls for, the rehabilitation of cunts through individual and collective acts that transgress both private and public boundaries (boycotting male-produced art and literature, tasting one's menstrual blood, retaliating against alleged sexual assailants). I understand at least one of Muscio's goals to be in line with my own, that is, a more integrated genital "lived experience" for women who have been shamed out of one for too long, the *clean space* postulated in this book. But Muscio's cunt is more—or at least differently—symbolic than mine, and our strategies for utopic incorporations are also distinct.

Rather than making the vulva manifest through confrontational and separatist political activism, I seek to delineate and analyze the practices, habits, and discourses through which genital alienation is produced and sustained. And, unlike Muscio, I am reluctant to prescribe remedies until I have made a more careful diagnosis. Muscio takes the sexual desires, confidence, and rebellious potential of women for granted, inviting them to simply get on board with her "cuntlovin" revolution (1998, 9). I have argued thus far that vulvar pain patients—who index the sizable proportion of US women that I believe suffer from genital dis-ease—are often neither interested in this nor able to do so; many of the patients I met explicitly longed for the day when their vulva could once again recede from their direct attention. Interviews with and fieldwork among these women muddy this "refusal," however, and reveal snapshots of ambivalent and confused desires. The culturally informed distance felt by many women toward their sexual bodies can be both mitigated and intensified with the emergence of pain; the attention demanded by their symptoms adds dimension—even if painful—to their previously missing genitalia. My interviews with these women are especially illuminating in that they were gathered from within this space, one that, despite its emotional discomfiture, is infused with a sense of possibility.

The Fragmented Self: Scout

Irigaray (1993a) argues that woman has been "torn apart" at the hands of masculine culture and that "she has never regained her wholeness" (114–15). Reluctant though I am to project a "whole" woman into a utopian past or alternate reality, the fragments and alienation that I describe in this chapter provide ethnographic evidence of the incompleteness of the contemporary female genital body. My interview with Scout, however, allowed me to glimpse some of the faraway places from which a "whole" female sexual body might be accessed. As we talked in her home one cold, rainy afternoon, she described a time when she briefly, and consciously, occupied a kind of sexual razor's edge:

> SCOUT: I can have multiple orgasms. And I remember one time that . . . that I mean, I just . . . He says that . . . I must have had twenty. But any[way]—who was counting (laughing), I don't know. But anyway, it was . . . I never forgot this moment because it was, I think sometimes, um, . . . I wonder what . . . *I hold back*. Because at that, that one experience for me was . . . Sometimes it's almost so pleasurable it's painful, and I don't want to know *what else I can find out*. And that day, I didn't care. And I was just amazed. So I know there's a lot of potential and a lot of sensations, but with stress and with *whatever else* . . . I . . . I . . . I **believe**, or, I don't know. I question whether *I hold back* . . . and whether I could experience *a lot more than I am* . . . and . . . Because I did.
>
> CHRISTINE: So what would happen . . . If you had to make up what it is that you might be afraid of, or what you're holding back from, what do you think it is?
>
> SCOUT: I don't know. I don't know. That's the most frustrating part. (Bold emphases are Scout's; italicized are mine.)

Scout had a fairly recalcitrant case of vulvodynia: only moderately responsive to oral and topical medications, it posed significant problems for her position as a security guard, where she spent a lot of time driving and sitting. She had been mildly reserved when I met her at the hospital, and she had agreed to an interview more out of a desire to tell her story than to help me with my work. But by the time we got to this point in our conversation, we had covered a good deal of intimate ground. Scout described an upbringing that was both strict and religious, and she spoke plainly about the toll that increasingly baggy and shapeless clothing was taking on her sexual self. Writing this, years later, I can recall the aforementioned exchange with crystal clarity, as if I were still sitting at that dining room table with Scout. When she spoke, I was completely caught up in her narrative, and I felt a space opening up on the table between us. When I posed my question, I was almost whispering, hoping I could both honor and preserve the story she was struggling to tell.

Scout's narrative evinces a tension between holding it together and falling apart: while intensely recalling the pleasure of the moment she describes, she simultaneously acknowledges the strain of accessing *even the idea* of what she suspects she might still be missing. Lynne Segal (1994) suggests that in a culture influenced by masculinist psychoanalysis, "a woman cannot exist except in the shadow of the phallus, which is what makes her sexuality so enigmatic" (132). She elaborates, addressing a "place of 'not-being'" defined by Lacan:

> "There is a *jouissance* [extreme pleasure] proper to her of which she herself may know nothing, except that she experiences it—that much she does know." But the unknowable truth of women's surplus of pleasure cannot, it seems, belong to the human sphere, to the symbolic world of the hegemonic phallus. It cannot, therefore, provide a woman with any way of communicating her existence as an agent of her own desire. (132)[5]

In this interview segment, Scout twice invokes the ineffable dimensions of her sexuality, wondering what *else* she might be able to erotically experience and "whatever *else*" (in her life) hampers her ability to do so. Many sexually active women, constituted as *lack* rather than unpredictable and generative *desire*, infrequently wonder if they are, indeed, "a lot more than" they seem to be. Culturally available sexualities address questions of feeling "better" either by trying to fix what is allegedly broken or by assisting women to procure as much penetrative sex as possible. Women with vulvar pain, in part through integrative PT, have the opportunity to bypass both these sets of interventions and explore what could be "better" in novel and productive ways. Potts (2002) argues that in "drawing upon conventional masculinist ideas of . . . sexuality, desire, and pleasures, . . . [women] may still be missing any alterity that might be associated with a recognition of female . . . desires" (208).

In her essay "The Three Genders," Irigaray (1993a) analyzes women's discursive sphere in reproductive terms:

> She has got lost in her role as mother, or else in a sexual display that does not really match her space of meeting and embracing. . . . Woman's value has been equated solely with her capacity to bear and to nurture a son, and to the language that corresponds to that function. (179)

Though I echo many of Irigaray's assertions, I am less inclined to conflate maternal and heterosexual discourses; I am especially disinclined to make this conflation given the remarkable lack of interest that several of my informants, particularly women with VVS, had toward motherhood. The task that Irigaray sets forth in this essay, however, of "discover[ing] and inhabit[ing] . . . the morphology of a [female] sexual body" (180), is one that, were I to construct it, would be near the top of my prescriptive agenda. And while Irigaray insists that the differences between males and females are irreducible, I, along with a host of gender studies scholars, prefer to delineate bodily difference in a nonbinary, or what Wilson calls "reticulated" (2004, 53–59), fashion, in a way that foregrounds my "attempt to see [difference] from multiple standpoints" (Clarke and Olesen 1999, 5; see also Barad 2007)). This means refraining from interpreting Scout's desire to know about her "potential," particularly one from which she might be "hold[ing] back," as reflecting an essential, knowable, or even recuperable female sexual identity. Rather, I locate her experience within a cascade of bodily possibilities, within what Potts (2002) refers to as "a proliferation of intensities over the libidinal surface, an 'opening up' of the senses to enjoy/participate in more than the visual or tactile—to spread over a diverse array of 'sensations' and 'happenings'"(239).

My alignment with Potts comes with an important reservation, however, in that the "body-less" sexualities over which she enthuses are confounded by the experiences of women with vulvar pain. Potts cites a list of influential thinkers—including Deleuze and Guattari, Lingis, Braidotti, Massumi, and, to a lesser degree, Grosz—who have influenced her ideas regarding radical female (hetero)sexuality. Conceiving of bodies as assemblages, as well as desiring-machines that surge with intensities and libidinal flows, these theorists "share . . . a 'vision for the future'" in which corporeality is "radically altered" (Potts 2002, 232). Each of them, she continues,

> posits a radical revision of bodies as erotogenic (libidinal) surfaces which are the sites of cultural inscription; each rejects the inevitability of oedipalized (phallicized) sexuality, seeking to eliminate teleology from erotic relations. They valorize difference and multiplicity, and posit desire as affirmative, rather than constituted by lack. (232)

Like Potts, I understand these perspectives to be critically sound and, indeed, cannot imagine theorizing extant and future sexualities (e.g., intersex, transgender, virtual, or technologically enhanced) without their insights. But my inclination to read as queer the nonnormative sexual behaviors of the women and couples that I met in the field was consistently thwarted by their explicitly stated desires for "normal sex." Unable to wholly incorporate heteronormativity, and therefore limited in their ability to transgress it from within (Beasley 2010, 208), women with genital pain held fast to the pieces that remained available, evincing the elusiveness of embracing sexual alterity while conformity still eludes one's grasp.

Feeling Female

Near the end of my fieldwork, Colleen came in to the clinic with a diagnosis of vulvar precancer (VIN I) that was acting a lot like vulvodynia.[6] Though she had already undergone a biopsy (which both excised and confirmed her condition), her vulva was still symptomatic, and, prior to her appointment with Dr. Robichaud, her case had been passed among several (male) physicians who could neither explain nor manage her persistent pain, redness, and skin fissures. "It's almost always raw," she reported, and "there are times it looks swollen."

> DR. ROBICHAUD: What happens when you have sex?
> COLLEEN: It hurts.
> DR. ROBICHAUD: [Is the pain the same or different from your] everyday pain?
> COLLEEN: Certain parts of the pain are the same; he tries to go into the pain.
> DR. ROBICHAUD: Are you able to have penetration?
> COLLEEN: Well, it's been hard. And then I feel pressure up in here. [Points to her belly.]
> DR. ROBICHAUD: Like something's pushing back?
> COLLEEN: Yes.
> DR. ROBICHAUD: Well, I'm going to have you participate in the exam. I'm going to have you show me. [The pain is your] body trying to prevent penetration. [Based on what you've told me so far,] you have everyday skin pain, pain with sex at the skin, and pain with the muscles with penetration. Anything else?
> COLLEEN: [It's] more sensitive around [my] . . . anus?
> DR. ROBICHAUD: How [are] your heart and your soul and your spirit holding up?
> [Nothing for a few seconds. And then Colleen starts to cry.]
> DR. ROBICHAUD: It seems like if you just opened up, you'd fall apart. You're so fragile.
> [My field notes say: "This patient is *really crying*."]
> DR. ROBICHAUD: Yeah, it's been really hard.
> COLLEEN: [There was a] battle [over] who would do [my] surgery [and I had a lot of fear about the cancer diagnosis. My] mom had vulvar cancer . . . and died of pancreatic cancer. [My] sister died of melanoma [and I am about to have a suspicious lesion excised from my leg]. But I [still] feel like a hypochondriac! But I can't ride a bike, I can't walk. And

I'm a waitress . . . !
DR. ROBICHAUD: How are things with your partner?
COLLEEN: He's been *great*! [Still crying a little.] [But] how do you *feel like a woman*? I can't *feel female*!
DR. ROBICHAUD: Have you asked him how he's felt?
COLLEEN: Well, he's more worried.
DR. ROBICHAUD: It's not cancer.
COLLEEN: Then how do we make it *better*? (All my emphases.)

Colleen told us that her vulvar skin often "cracked" and bled, and that two physicians had "made comments about the tissue being rice-paper thin." Dr. Robichaud was attuned to the clinical relevance of this fact; she also recognized the importance of assessing the fragility of another part of Colleen's genital experience. Like many of the women I interviewed, Colleen struggled with how to "feel female" when she could not participate in penetrative coitus, a "praxis" that Kaler (2006) refers to as the social relation that "link[s] heterosexuality to gender identity" (51). And though this construction can be explored from a number of angles, I want to examine how Colleen's "falling apart" (during her conversation with Dr. Robichaud) relates to the bodily and sexual disintegration that her pain had both catalyzed and begun to reveal. "Rather than looking to . . . the presence of previous pathology . . . to explain severity . . . of symptoms," argues Brown (1995), "we might begin instead to ask how many layers of trauma are being peeled off by what appears to be only one . . . event or process" (110).

Pain like Colleen's engenders a particular brand of corporeal alienation. It threatens bodily integrity because of "the way [it] enters into our midst . . . at once something that cannot be denied and something that cannot be confirmed" (Scarry 1985, 13). But the genital isolation produced by pain's role as "alien . . . intruder, [and] invader" (Jackson 1994, 209) is layered on top of a deeper and far more pernicious breed of isolation and alienation, in that the absence and erasure of female nonreproductive genitalia from the social landscape undermines women's ability to "have" their bodies in *any* form of entirety—painful or otherwise. These erasures are unremarkable and routine, what Brown (1995) refers to as the "continuing background noise" (103) of women's lived sexualities. The censorship of the vulva from allegedly polite conversation, its discursive and material contamination in mainstream heteronormative pornographic media, and the utter inconsequence that leads to the sanctioned cosmetic removal of so-called "redundant" labia (Goodman et al. 2010; Mayer et al. 2011) are everyday practices that pull the rug out from an always-already precarious genital integrity.

In describing the effects on all women of what many feminists label rape culture, Maria Root (1992) theorizes a form of "insidious trauma" as "the traumatogenic effects of oppression that are not necessarily overtly violent or threatening to bodily well-being at the given moment but that do violence to the soul and spirit" (in Brown 1995, 107). Like Judy (the sixty-two-year-old woman whose fused vaginal opening I describe in the previous chapter), Colleen's clinical encounters before she reached Dr. Robichaud were riddled with an affect perhaps best described as nonchalant. Indeed, Colleen's story of being tossed like a hot potato among physicians for whom her symptoms were just *too much* reflects the kind of casual disregard for her genitalia through which it is possible to understand the emergence of cosmetic vulvar surgeries. Only in response to this kind of inconsequence could Judy find her physician's suggestion to let her LP-afflicted vulva remain fused even remotely reasonable. "'Real' trauma," suggests Brown (1995),

> is often only that form of trauma in which the dominant group can participate as a victim rather than as the perpetrator.... The private, secret, insidious traumas to which a feminist analysis draws attention are more often than not those events in which the dominant culture and its forms and institutions are expressed and perpetuated. (102)

Vulvar disregard is normative and, in these terms, "outside the range" of hegemonic definitions of trauma. Brown's (and my) goal is to widen the experiential range against which traumatic events are measured in order to include the "experiences to which women accommodate; potentials for which women make room in their lives and their psyches" (101). This is because women in the contemporary United States, though not (yet) traumatized by the specter of vulvar pain, have nevertheless "made room" for a reality in which their genitals often matter little (Moore and Clarke 1995).

Capacities for Action

In her seminal work on corporeal feminism, *Volatile Bodies*, Grosz (1994) contrasts a condition known as agnosia with that of the "phantom limb," that is, sensations and pain that persist in the space of a surgical amputation or other form of loss: "Agnosia is the nonrecognition of a body part that should occupy a position within the body image. In traditional psychological and physiological terms . . . [it] is seen as a forgetfulness, a refusal or negative judgment" (89). Grosz's descriptions of these neuropsychological states are situated within a larger discussion of corporeal phenomenology, in which she points out that Merleau-Ponty understood both phantom limb and agnosia to "demonstrate a fundamental ambivalence on the part of the subject," as, in the case of the phantom limb, "actions which the arm, say, would or could have performed are still retained as possible" (89).

But though both conditions involve ambivalence, it is important to distinguish that the phantom retains a potential for action that the agnosic body does not. Constructed and beheld by discourses of disavowal, women with vulvas live out a deep and diffuse genital agnosia that stifles and paralyzes their ability to "have" their genitals in meaningful ways. These corporeal phenomena are bound up within the *body image*—a postural and proprioceptive map, or "schema," that "registers current sensations . . . preserves a record of past impressions and experiences, [and] . . . is formed out of the various modes of contact the subject has with its environment through its actions in the world" (Grosz 1994, 66–67). The body schema, Grosz continues, "is an anticipatory plan of (future) action in which a knowledge of the body's current position and *capacities for action* must be registered" (67; my emphasis; see also Sobchack

2010). The body image—overdetermined by a combination of physical, cultural, and intersubjective experiences—is multiply informed, continually produced and invested, and susceptible to the dominant cultural norms that influence its social value (Schilder 1950). This profoundly cultural product can help us make critical distinctions between the kinds of "ambivalence" expressed in these two conditions; that is, the fact that different bodies, and different *body parts*, are accorded varying amounts of cultural capital is related to the fact that they maintain distinct capacities for action.

Feminist analysis requires an attention to the gendered aspects of these differences in body image and capacity. For Brown (1995), such analyses "illuminat[e] the realities of women's lives, turn[ing] a spotlight on the subtle manifestations" (108) of lived difference. For Grosz, it is the physical peculiarities associated with sexual difference that must not be neglected when examining these body maps:

> It seems incontestable that the type of genitals and secondary sexual characteristics one has (or will have) must play a major role in the type of body image one has and that the type of self-conception one has is directly linked to the social meaning and value of the sexed body. (1994, 58)

The words of Brown and Grosz allude to the long-held feminist tenet that women are culturally constructed as lacking, missing, and damaged, and that this social fact has long influenced women's distinct ways of being-in-the-world. And though recent transgender and intersex scholarship has begun to (rightly) critique and reformulate such models of sexual difference (Salamon 2010; Spade 2006; Valentine 2012), I maintain that the vulva—through the sustained, routine, and unquestioned deployment of censorship and shame—acquires a unique dimension of bodily absence. Missing from even a somewhat recent (and relative) explosion in "vaginal" popular culture (Armstrong 2010; Ensler 2001; Fabello 2013; Wolf 2012), still too much for the masses, the morphology of the vulva is constructed by collective disavowal and absorbed by individual bodies.

Vulvar Dys-appearances

Some bodies, confronted with inexplicable and sexually prohibitive pain, are compelled to take action in the service of locating relief. But how does one take individual action on behalf of a collectively ignored part of the body? How and where does one identify and locate the source of one's pain if that source did not meaningfully exist until it began to hurt (Leder 1990)? How does one attribute corporeal agency to genitals from which nothing has ever been asked or expected? How does a vulva become a "*knowing*" body part, one "that shares in the knowledge of the body" (De Preester and Tsakiris 2009, 309–10)?

One afternoon, in an interview over lunch at Denny's, I asked Ashley about where vulvar discourse most appropriately belonged.

> ASHLEY: Sex ed. I mean, you know, if they're gonna teach about all, everything else, then they really need to teach about that. And, in a way, excluding that makes it . . . off-limits. Something **we** don't talk about. We certainly talk about penises enough. Anything that a guy might get wrong with **them**, we all know all about. You know, I can't think of anything that we haven't . . . But, if it goes wrong with a woman, that's just not acceptable.
> CHRISTINE: How much do female genitals matter?
> ASHLEY: Well, as long as they're *usable*, I don't think they care much. [laughs] Um, I think if they're **not**, I think you'd be hard-pressed to find a date for the prom! [laughs] (Bold emphases are Ashley's; italicized are mine.)

Vulvar agnosia is entangled with the investments that others hold in labial stability, in the seamless and silent role they are meant to play in vaginal penetration. In other words, a vulva doesn't/isn't really "matter" (Butler 1993) until its disability obviates activities through which many individuals generate their sexual subjectivity. Irigaray (1992) transforms this silent and "usable" flesh into abundant and contiguous lips that offer women the pleasures and integrity of "self-affection" (Labuski 2008). In resignifying female genital lack as a proliferative body that "goes on touching itself indefinitely, from the inside" (15), Irigaray makes the vulva active, generative, and multiple, endowing it with what Vivian Sobchack (2010) calls the "motor capacit[y] to fulfill a present intention" (62). But the masculinist ideologies that dominate the practices of heterosexuality and medicine do not have room for these dynamic lips. I asked Ashley to elaborate on what would make a *woman* unusable, and she told me about a friend she knew who had "lost her vagina" to cancer, presumably through pelvic/genital irradiation. This notion of a "lost vagina" is a provocative one, allowing us to question whether and to what degree an absence, or a "hole," can be rendered impotent.

For (male) partners and (many) physicians, the world proceeds more smoothly if the vulva quietly recedes, much like the visceral organs and physiological systems described by Leder (1990) in *The Absent Body*. Leder elucidates these unquestioned recessions—the livers, glands, and kidneys of which we do not take conscious note, even as their activity assures our day-to-day vitality—and argues that "these movements are not experienced as within the 'I can' of personal mastery. I do not feel a sense of guiding or controlling these processes" (46). But unlike these viscera, the vulva is external anatomy and need never recede from a woman's conscious body image. For symptomatic women, therefore, the difficulty in delineating the specifics of *down there* indexes a maladaptive *dys*-appearance (Leder 1990; Manderson 2011), an alienation marked by the inappropriate or uncomfortable appearance of an otherwise quiescent body part, mystified further by various—and culturally sanctioned—transfers of ownership of female genitalia. Increasingly aware of the integral role in their husband's sexual satisfaction that their vulvas were scripted to play, clinic patients often spoke of their appointments in terms of *his* frustrated investment: "I wouldn't even be here if it wasn't for him," for example, and "This wouldn't bother me at all if I wasn't married." Similarly conscious of their doctor's interest in compliant genitalia, they repeatedly apologized for their lack of formal knowledge, as well as for their compromised ability to tolerate routine gynecological attention.[7]

Afflicted women's bodily confidence was riddled with insecurities. Rosemary, who at eighty-two was the oldest woman I met during my fieldwork, greeted Dr. Erlich with tremendous clarity, stating simply: "I heard about the vulvar clinic and [I] thought you might be able to tell me what's wrong." But as Dr. Erlich gathered her history (which included a "fairly sexually active" marriage), Rosemary felt the need to humorously address the assumptions she feared were being made about her. "I feel a little embarrassed," she said, "[because] I know you're really busy. [But this is] really not in my head. I'm not a hypochondriac. If I was, I would have picked a different symptom."

In the years that I practiced as a clinician, I saw and medically managed thousands of women whose sexual bodies reflected the genital agnosia that I suggest is culturally hegemonic. From a perceived inability to remove a forgotten tampon, due to a poor understanding of their vaginal anatomies (risking serious infection), to active aversions toward contraceptive methods that required genital contact or manipulation, women repeatedly revealed their preference to just not *deal with* their genital bodies. These vulvar *dys*-appearances index a body image mediated by an actively disparaging world, one that circumscribes women's ability to live their genitalia in an inconspicuous fashion (De Preester and Tsakiris 2009, 311).

The forgotten tampon, like an unrecognized sexually transmitted disease or the early symptoms of vulvar cancer, reminds us that genital alienation can indeed have permanent bodily consequences. But the patient whose choice of pills over a diaphragm does not necessarily entail a greater risk for bodily harm still participates in the partial (and routine) excision of her vulva; in this case, by delinking her genitalia from the contraception that enables her to engage in nonprocreative coitus.

I suggest that even this partial excision is worth heeding, as it articulates with numerous and cumulative other sites of cultural erasure, few of which are purely symbolic; rather, they show up in the "muscles, nerves, and organs" (Wilson 2004, 5) of the bodies vulnerable to their effects. Echoing Eve Ensler (2001), I'm "worried about" (3) vulvas. I'm worried about the disappearing act in which US women appear to have them engaged, as it neither lends itself to a female-centered genitality nor benignly skims the surfaces of the bodies in question.

Precariously owned and minimally recognized, the underintegrated genitals of symptomatic women make themselves manifest, rupturing the silent seamlessness through which they have thus far been lived. But the stories I heard at the vulvar clinic suggest that discourses of inconsequence and erasure are powerful enough to staunch the effects of these eruptions. Describing their pain to ears unattuned to their vulvar well-being ("Well, you've got lube, so you're fine!"), annually insisting that things were sexually "not right" but being met with the examined opinion of their gynecologists that there was simply "nothing wrong," and being routinely cajoled by allegedly normal friends that they should just "move through the pain," clinic patients struggled to make sense of the cultural and medical narratives that were at odds with their bodily experience. In this sense, their psychic and cognitive dissonance resembled those of an amputee with the first tingling sensations of a phantom limb: wondering if they were crazy to feel something (where) they were evidently not supposed to.

TOWARD A VULVAR-BASED SEXUALITY? ASKING NEW QUESTIONS

Naturalizing Absence: Brigette and Sharon

As a concept, genital alienation has both depth and tenacity: it presumes a body immersed in and physically obliged to its cultural milieu, and it allows us to think critically about what is physically *present* in a body part rendered *absent* at every turn. This absence pervades the female sexual body image and severely restricts women's abilities to generate desire and sexuality on their own terms. The issue of how to "feel female" with which Colleen was explicitly struggling arose repeatedly in my interviews. I asked my respondents how much of their sexuality they believed to be bound up with—or defined by—their genitalia, and they often answered much like Brigette: "As a woman? Oh, I'd say probably like ninety-five percent. I mean, it's a *huuuge* part, especially in society nowadays. I mean . . . [long pause] Yeah, I'd say it's a large part of being a woman."

Brigette also told me that sex with her current partner was often disrupted by her pain, but that this did not cause an inordinate amount of conflict between them. Her previous partner, with whom she was involved during the onset of her symptoms, "didn't have a clue" what she was talking about when she initially broached the subject, and so she kept the severity of her symptoms from him for the remainder of their relationship. Although Brigette's current partner, Jim,[8] who joined us for the interview, was as "curious" as she was about the body (he was pursuing a career as a mortician), and although they both "like[d] having sex a *lot*," Brigette continued to gauge her symptoms against a standard of penetrative intercourse. Her discomfort, she said, "didn't really matter" outside of a (hetero)sexually active relationship; this was in spite of her complaint that "you can't study in school when you're feeling all . . . like, you know . . . in pain or itchy or whatever," and that she had had "about seven biopsies" thus far on her vulvar skin.

Brigette's quite typical perspective on the symbolic value of her genitals was missing the kind of (female) carnality described by both Butler (1993) and Wilson (2004), the "mattering forth" (Povinelli 2006, 7) of an integrated body that speaks both from and for itself. In the absence of one's own genitals, sexual identity is most sensibly defined through the terms of another person; having penetrative intercourse (ever or once again) promises the closest thing to "feel[ing] female" that many heterosexual women have experienced (Kaler 2006). Sharon, whose lichen planus was making her husband's typical sexual behaviors very uncomfortable for her, expressed it this way:

SHARON: I think if I weren't in a relationship, I think I could become very dist—, distant with this . . . thing right now, and just say, "Well . . ."

CHRISTINE: And maybe not be doing the dilator therapy?

SHARON: Right mmm-hmm. Kind of . . . distance myself from that, the sexual part of myself, and say, "Oh well!" [laughs]

In a body image from which a substantial portion of a "sexual self" (Ogden 1999, 2003) may be neurologically or psychologically absent, the capacity to act in a generative manner is severely restricted. As I have discussed, relinquishing the desire for normative intercourse is a step that bodies alienated by pain or agnosia are perhaps not ready to take, and scholars attempting to "queer" female heterosexuality have thus far both underestimated the inscriptive and infiltrative power that dis-eased cultural discourses wield over the process of genital integration and overestimated the freedom and "capacity for action" sustained by agnosic genital bodies.

To be sure, women in Western-influenced cultures and locations are the undisputed beneficiaries of a host of civil rights, along with feminist and sexual revolutions that brought matters of the sexual body to the political and discursive tables; many women undoubtedly enjoy the sexual liberties that accompany such profound cultural shifts. Indeed, both popular and alternative media are rife with (hetero)sexually *active* female bodies that, in Lynne Segal's words, "have everything to gain from asserting [their] non-coercive desire to fuck if, when, how, and as [they] choose" (1994, 314; also see Beasley, Brook, and Holmes 2012 and Vance 1993). And while I applaud this enthusiasm, my encounters with female patients—whether as nurse practitioner or anthropologist—resonate more profoundly with the words of Merri Lee Johnson (2002), who insists instead that "while feminism may have freed women to fuck, the fuck—and the 'role of the fuck in controlling women'—has in many ways stayed the same" (23; see also Cacchioni 2015 and Fahs 2014).[9] Ethnographic attention to the desires and proclivities of women with vulvodynia and VVS—bodies that are behaviorally queered and empowered (by pain) to refuse heteronormative penetration (Labuski 2014)—reveals, instead, that feminists may have not fully acknowledged the bodily integrity upon which an authentic, female-driven set of sexual choices ultimately needs to rest. For Stevi Jackson (2006), heteronormativity regulates the behavior of those both outside *and inside* its borders, and the "opportunities to escape its most conventional forms are not equally distributed" (112).

In their research on first intercourse among teens in the United Kingdom, Holland et al. (1998) came to similar conclusions, that is, that apparent complicity with available sexualities does not necessarily reflect an achieved sexual parity between heterosexual women and men. Even women who refuse male-centered intercourse for reasons other than prohibitive pain are, according to their analyses, "trapped in their resistance, finding it easier to disrupt male definitions of desire and natural dominance than to *produce female desire itself*" (in Potts 2002, 208; my emphasis). Irigaray foreshadows this lament in "The Three Genders," stating: "As for woman, her moves as a lover have still to be invented" (1993a, 179). I return to the production of desire in the following chapter, but here I want to highlight the disjointed and unintegrated aspects of this sexual state of affairs, as well as the ways that vulvar and genital agnosia are at the heart of this disability.

In our extended interview, Brigette drew parallels between the socially awkward natures of her vulvodynia and the (also transgressive) bipolar disorder with which she was diagnosed in her early teens (she was twenty when we met). Although Brigette said that she talked openly with friends about her well-controlled "manic-depressi[on]," she was more reluctant to discuss the genital symptoms that occupied far more of her time and attention. When I asked her to elaborate, she said, "Why don't I talk about it? I don't know. It just usually doesn't even come up in conversation." Brigette herself was nonchalant about this disparity; it worked for her in the immediate sense because she predicted that her friends would "probably be uncomfortable" with the topic. But the lingering and verbal absence of Brigette's vulva—in pain or not—from conversations with friends indexes the reticent and contaminating habitus in which the female genital-sexual body exists. This absence seems particularly acute given that Brigette talked openly about "just having sex or whatever" with these same friends.

On my prompt, Brigette compared and contrasted her dual diagnoses. She first stated that she understood her bipolar disorder to carry a "worse stigma" than her vulvar pain, but then she complicated this assertion by reflecting that they both led her to feel—at times and differently—"out of control": "They are very similar. It's like, if I were to compare any two parts of my life, I wouldn't compare, like, you know, mental health problems to like, arthritis, but I would sa— I would compare these. They are very similar stigmas." Brigette's lack of concern or frustration over this routine (self-)censorship is powerful evidence of an arbitrary linguistic proscription that, though it has become naturalized (Bourdieu 1984), nevertheless registers an experiential impact. Brigette knows that the topic of her diseased vulva would in all likelihood make her friends uncomfortable, without ever having to confirm this. But what is most critical to notice is that it is *only* her pain that would have led her to bring up her genitals at all. Ever. Without pain to animate them, female external genitalia remain lifeless, in an ongoing relationship with the social conditions of repression and invisibility.

That *Kind of Sex*

In theorizing female sexual desire, we must account for the disintegrated, silent, and agnosic bodies that index Vulvar Disease. Postmodern and queer erotic imaginaries must reconcile the surface and extragenital sexualities they posit with the lived experience of women who have not (yet) participated in heterosexuality's most normative event—its signature gesture, if you will. VHC patients convincingly conveyed this reality. Even I, who had gone to the field almost certain that vulvar pain was the embodied opportunity to "move on" from phallocentric coital imperatives, found myself rooting for the success of patients' surgical, pharmacological, or homeopathic efforts toward having the kind of sex that mattered to them. Isabelle, who had been symptomatic for four years when she flew from Atlanta to consult with Dr. Erlich for her VVS, and who had not yet "consummated" her marriage of three years, was particularly articulate about the meaning that penetrative sex held for her:

> ISABELLE: I think I've always kind of associated, uh, vaginal intercourse with, um, a more spiritual connection, you know? And I think that's one thing that . . . not, it's not all about [the] physical, but it's more um, you know, just . . .
>
> CHRISTINE: Can you tell me more about that?
>
> ISABELLE: Yeah, well um, I think it comes from uh, my . . . my faith, um, mainly. Uh you know, just, um, I think (of course since I've never had sex I don't know), but there's just, uh, the emphasis that, uh, Scripture puts on . . . sex, you know, um, uh . . .
>
> CHRISTINE: On *that* kind of sex.
>
> ISABELLE: Yeah, *that* kind of sex. Um, that, uh, you know, even if, really, kind of, the way I've, . . . My philosophy or whatever is that even couples who are not married, who um, who have . . . intercourse, are kind of joined, you know? And so, um, that's kind of just uh, . . . [the] way I think of it. More sacred. Not necessarily better or anything, just, uh . . . uh . . . *the most intimate*, I think, um, way that a man and a woman can be joined.
>
> CHRISTINE: And something that you still very much want.
>
> ISABELLE: Yeah.
>
> CHRISTINE: When you think about being able to do that with your husband, what is, I mean . . . How, what is it like to think about finally being able to do that?
>
> ISABELLE: Yeah, it, I think just uh, just relief. And, uh, you know, I think . . . a physical closeness we've . . . not been able to have, uh, you know. We're very intimate, very open, and, uh, have fun. I mean, we're creative and so it's not that, uh, I mean, even if we never could, it wouldn't um . . . uh . . . you know,

we have a . . . It's not that I feel like we don't have a real marriage, but um . . . Uh . . . it would be something missing. I can't, you know, I won't say that I wouldn't feel that way, but uh . . . But, [it's] just because I think it's, that's the way we're created to be, as man and wife. Ultimately.

When I asked Isabelle if her ideas about intercourse as a sacred joining were related to beliefs about conception and reproduction, she unsettled some of my own assumptions by thoughtfully responding, "To be honest, I really don't think I want children, so I definitely don't associate it." For Isabelle, then, the practice of penile-vaginal intercourse indexed neither a naïve adoption of cultural norms (i.e., an obligation to bear children) nor a rote submission to the procreative prescriptions of a conservative religion (Carpenter 2005). Rather, being able to have "that kind of sex" was deeply symbolic of the spiritual commitment she and her husband had made to one another, something from which she might never be interested in moving beyond. Isabelle's narrative challenges my own feminist thinking about religion and gender and reminds me that the apparently simple *practice* of normative behavior most often reflects complex negotiations between the "three bodies" under consideration here—individual, social, and political (Scheper-Hughes and Lock 1987). Regardless of my personal politics about Christianity's relationship to female sexuality, and of our disparate ideas about the definition of "real sex," women like Isabelle—doubly alienated from their genitals (through pain and erasure)—showed me that owning heteronormativity is a prerequisite to giving it away.

In the contemporary United States, "the fuck" is the most direct route to the cultural capital promised and enjoyed by members of dominant (i.e., straight) society. Critical feminists like Segal, Grosz, and Potts must acknowledge that refusing or displacing it on ideological grounds is an act that fully depends on the privileged ability to accept its "pleasurable possibilities" (Beasley 2010, 205). If the "goal of change," argues Teresa Ebert, is "the removal of all restrictions, all limits on the *play* and *pleasures* of sexual differences" (1996, 163; emphasis in original), then we must attend to the totality of these possible differences, as well as to their material bases. Focusing on the extant bodily differences *among* women (Crenshaw 1991; Mohanty 1991), leads me, therefore, to offer the following caution: if calls to "enact the erotic in politically alternative ways" (Potts 2002, 255) are framed through a feminist lens, then those "alternatives" can exclude neither my informants nor any other woman marginalized from the routine enjoyment of sexual normativity (Cerankowski and Milks 2014). Strategies of "deessentializing" should not, as Beasley (2010) argues, fail to account for the "asymmetric constraints in existing social relations [nor] the constraints of visceral physicality" that constitute the "social flesh" of many women (208).

Indeed, symptomatic women are strangely situated with regard to the politics of dominant heterosexuality: with access to its practices proscribed by pain, the privileges associated with those practices slip more and more easily through their fingers. The cultural cues that index a bodily familiarity with conventional coitus are less available, compounding their shame-based isolation with the experience of being chronically *out of the loop*. The formation of a resistant (and potentially empowered) subculture, however, is also unavailable, as the majority of these women are ideologically aligned with this same dominant sphere (Hebdige 1979). The discursive transpositions implicit in the "queering" of heterosexuality, then, seriously belie the distinct experiential worlds in which dominant and subordinate bodies exist. Caught up in the "exemplary normalization" (Beasley 2010, 208) of heterosexuality, few of these women have cultivated resources that could engender radical—and potentially liberating—sexual subjectivities.

The danger of being in the majority, of your life being positioned in the mainstream, is that your poor fit—your *nonintegration*—cannot be easily disguised from your peers. When conversations with other women turned to (hetero)sexual matters, for example, patients found themselves profoundly alienated and sometimes "freaked . . . out," in Isabelle's words. Clair, whom I have previously introduced, and who could not tolerate the same-sex intimacy of PT on religious and sexually conservative grounds, found that her VVS undermined her access to female friendships:

> Um, . . . I teach seminars and workshops on friendship, so I'm really very . . . um, I, I know the extent of women having relationships and friendships is really important. But yet, . . . when a group of women are sittin' around and one of 'em says, "Oh my gosh, we have the best sex," this wall goes up and I think, "I can't relate to you." And yet, . . . it's also there's that embarrassment. There's that "Oh, if they only knew that I haven't had sex in five years," they'd say, "Oh you poor thing," you know. *That*, I think, that may be, um . . . maybe that's a part of it. Not that I have a lot of women that do that, but . . . but I *envy* that. When I hear other women talking, there's this, and I go home thinking, "Uh, gosh I envy that."

Both Isabelle and Clair poignantly narrate a subjective reality shared not only with their fellow patients but with any number of women whose exposure to *Cosmopolitan* magazine, *Sex and the City*, or online pornography leaves them feeling as if they are sexually missing out on something. Women with chronic and disruptive vulvar pain, however, are estranged from both ends of the ideological spectrum in distinct and peculiar ways. Falling through a host of cultural cracks, their integration into the (hetero)sexual worlds around them remains as fragile as their vulvar and vestibular tissue.

Critical Juxtapositions: Vulvar Disease

The grinding halt to which symptomatic women's lives often come functions as a kind of threshold between bodily and social alienation and a thus far unrealized corporeal *home*. Having to put their sexual desires on hold allows these women the opportunity to critically examine the assumptions and hierarchies around which their desires might otherwise be organized, to question what *else* might structure or inform the sex they hope to eventually enjoy. Their successful transition over this threshold is dependent on a set of factors linked to their individual, social, and political bodies—specifically, on the ability of these bodies to meaningfully cohere. At a minimum, these factors include: recognizing the abnormality of symptoms, locating a knowledgeable provider, having the resources to obtain services (e.g., insurance, transportation, time), trusting the provider and reporting an adequately detailed medical history, being in at least one supportive relationship, having the discipline to comply with complex treatment regimens and the physical ability to tolerate them, balancing a combination of providers and therapies, and accepting the possibility that some pain will always remain.

In my experience at the clinic, I observed almost one hundred women grapple with some combination of these facets of their disease, many of which are faced by people with other chronic illnesses and pain conditions. Vulvar pain is peculiar, however, in that the presence of its most frequent comorbid condition—pelvic floor myalgia—is directly related to the ways that a woman has thus far coped with her symptoms. Given the negative symbolic value of her genitals and their suffering, this coping has almost always involved hiding, repressing, and censoring her condition from all but the most invested of actors, making myalgia, at least partially, an epiphenomenon of the cumulative effects of Vulvar Disease. Acknowledging the reality of this syndrome constitutes an eruption, and symptomatic women may either come together or fall apart in the wake of doing so. As they attempt to reconcile their now pain-filled sexual lives, these patients must face not only the accumulated baggage with which their symptoms are culturally bound but also the possibly stolen futures of sexual satisfaction and fulfillment, futures that are as profoundly shaped by social worlds as they are by personal desire.

6

GENERATION

Novel Morphologies

CHRISTINE: What do you think this experience has given you?
ISABELLE: I guess in having to talk about my problems so much, I have been able to actually help a few women I've known, you know, with other problems. Well, like [when] my sister . . . discussed some problems that she had and that she was so . . . I mean [she] could not talk to her doctor about them at all, I mean, *could not even talk to her doctor.* And here I am having to tell my doctor everything, you know, right down to the most intimate detail.

—From my field notes

Her thigh is in her head, her mind is muscular.
—Elizabeth Wilson, *Psychosomatic*

REORGANIZING THE BODY

Success

When Daphne called me in September 2005 to tell me she had recently "had sex" with her boyfriend for the first time and that *it hadn't hurt,* we were both ecstatic. We felt especially good about this event because it was Daphne's first time having intercourse with *anyone.* When I left the field, just a month prior to our phone call, she and Brandon had been engaging in lots of heavy petting and genital contact, and

Daphne was working steadily with Hanna in physical therapy (PT). Indeed, it was during the last PT session for which I was present that Daphne inserted her finger into her vagina for the first time (also without pain), an experience that helped her to establish a corporeally meaningful difference between what she referred to as her "inside" and her "outside." I have argued elsewhere (Labuski 2008) that Daphne's painless intercourse was fundamentally enabled by her ability to preserve something of herself during sex with her boyfriend, a process Irigaray (2004) refers to as self-affectionate virginity. Daphne's intercourse with her boyfriend was also created by—and generative of—a novel *body image*, one that was informed by both an acceptance of her pain condition and a curiosity about the degree to which that pain would continue to figure in her (sexual) life.

Daphne both imagined and engaged in sex with Brandon through a body image that she constructed with Hanna, one that extended beyond her genitals and her sexuality. Not long before my last session with her, Daphne arrived menstruating to one of her appointments and not particularly interested in doing vaginal work. But Hanna quickly made it clear to Daphne that working with her ideas, thoughts, and beliefs about her bodily potential was simply another kind of "internal work," and she reminded Daphne that all PT was in dialogue not only with the rest of her body but with her entire life. And so on this particular afternoon, instead of lying on her back on a treatment table or sitting in a recliner wired to the monitor, Daphne worked with Hanna from her feet, in an Aikido exercise meant to unsettle and then reconfigure the elements of her body image. Standing upright in the center of the room, Daphne closed her eyes and allowed Hanna to physically push at her from all sides so that she could practice "holding [her] ground" in bodily situations over which she had limited control. The two women conducted this exercise for about fifteen minutes, and Hanna never ceased to connect it with the more local/genital techniques in which they normally engaged. I watched in fascination as the contours of Daphne's vulva and pelvic floor became more expansive, eventually coming to include her entire body and its ability to tolerate being pushed around.

This "opening out" of Daphne's body is noteworthy, particularly as it contrasts with the "shutting down" described by so many of my informants. For Merleau-Ponty (1962), bodily loss and generativity are "thoroughly relational and co-constitutive" (Blackman 2010, 7); corporeality cannot be understood outside the affiliative associations in which bodies and persons are always-already immersed—with others, objects, and the world at large:

> Precisely because my body can shut itself off from the world, it is also what opens me out upon the world and places me in a situation there. . . . The . . . [loss] is recovered when the body once more opens itself to others or to the past, when it opens the way to co-existence and . . . acquires significance beyond itself. (Merleau-Ponty 1962, 165)

Merleau-Ponty uses a case of aphonia (vocal paralysis) in a young girl angry with her parents to illustrate that what we perceive as loss can also be understood as rejection—"as one loses a memory . . . . it is lost . . . in so far as it belongs to an area of my life which I reject" (1962, 161–62). And though I have questioned the degree to which many women even "have" their vulvas, the bodily tension proposed by Merleau-Ponty—between having, rejecting, and losing—is a compelling way to think through this dilemma. In relationship to a world that does not account for their genitalia, women like Daphne shut down, "lose," and psychically reject the sexual bodies indexed by their genitalia, keeping their vulvas at (a sometimes literal) arm's length with behaviors that are only partially voluntary.[1]

With Hanna's guidance, Daphne was able to recuperate her losses as she lay claim to a set of new corporeal connections. For example, on the day that Daphne inserted her finger into her vagina, Hanna made certain that Daphne took full account of *all* the parts of her body that were helping to keep her pelvic floor in a state of relaxation, including the musculature of her jaw and abdominal diaphragm. And just as Lisa had asked Nikki to name and speak with her past (see chapter 4), Hanna encouraged Daphne to give her future a name:

HANNA: [Choose] a word that means relaxation to you. This is where soft is. . . . Now, while you're relaxed, I'm going to add your breath [to the monitor]. . . . When [you] need that soft quiet place, [think,] "I know at least two ways to help me get there." Go inside and think, "This is it, this is what it will feel like." Now close your eyes, mess your body up a little, and then see if you can come back to this, [if you can] call it up.
[Daphne does so almost immediately.]
HANNA: That's lovely. You get an "A" for today. Can I ask what word you used?
DAPHNE: Europe. [I went there two years ago and I] never felt better. [This was related to a man I knew there, with whom I] felt safe. Very safe. Europe [evokes that feeling].
HANNA: Okay, you need to think sometimes, "My body knows how to do Europe. I need to be in Europe." Even when things are ugly.

For Merleau-Ponty, Veronica (the name for Nikki's pain) and Europe exist as states of recovery, collected with a body that can "once more" open itself to a rejected, forgotten past. I want to complement this assertion and suggest that we understand the work of both Nikki and Daphne in reconfigured temporal terms. That is, as *future* possibilities through which patients can imagine a different set of corporeal experiences (e.g., sex without pain), Europe and Veronica exist as both recovered relationships *and* as novel morphologies, both accessed through a somatic mode of attention that is specific to vulvar pain.

In this chapter, I theorize two kinds of generation: the bodies created and produced through physical therapy sessions like Daphne's and the behaviors and ways of life that women cultivate through their experiences with vulvar pain. In the first section, I use concepts from phenomenology and contemporary neuropsychology to describe the alternative temporalities and embodied multiplicities developed by some women in physical therapy; I then use Foucault's notion of generative power to interpret vulvar pain's productive capacities. Finally, I use transcripts from interviews, during which I asked patients to discuss their pain in terms of gains and losses, in order to demonstrate the categorical elusiveness of a condition marked by its ability to render the absent present.

New and Renewed

For Merleau-Ponty, bodily integrity is achieved through restored connections. His predilection toward a body capable of psychically repressing experience is explicitly influenced by psychoanalytic theory, and, like Freud, he signifies the importance of an *event* as both treatment modality and evidence of its necessity. In the case of the young girl with aphonia, Merleau-Ponty suggests that "the girl will recover her voice, not by an intellectual effort or by an abstract decree of the will, but through a conversion in which the whole of her body makes a concentrated effort in the form of a genuine gesture" (165). Repudiating the idea that the mind alone is capable of bringing about a desired and therapeutic change, such as recovering a voice or a relaxed pelvic floor, Merleau-Ponty instead underscores the physicality and the "genuine-ness" of a gesture. But his reliance on psychoanalytic tenets is revealed not only by how he theorizes psychosomatic manifestations but also by the certainty through which he locates both cause and cure in an afflicted body's past.

I have suggested that acknowledging and recovering "the past" plays a crucial role in a total confrontation with vulvar pain. But these pasts can be both collective and cultural, and working them into clinicians' "etiological pathways" requires a gaze that looks both at and beyond individualized, biologized, and narrowly construed criteria for "other causes of genital discomfort" (Harlow and Stewart 2003, 93). With critical and ethnographic participant observation (i.e., moving within and across diagnostic and treatment realms and making room for multiple interpretive frameworks while collecting patient narratives), I was able to more fully appreciate the respective roles of the pasts as well as the *imagined futures* in which vulvar pain might be resolved or reconfigured. Treatment approaches that focus on "curing" *vulvodynia* or VVS by excising or anesthetizing inexplicably damaged tissue or by systemically altering a single physiological element effectively "shut down" the patient's experience of her symptoms, anatomically and temporally constraining their meanings.

Although constituted within the biomedical discourse of "a future without pain," treatment options such as surgery and topical anesthesia fix a symptomatic woman's experience into an imagined past, one in which sex was not painful. Rather than envisioning how they will move forward into new relationships with their (genital) bodies, both patients and physicians construct narratives that focus on "never having to think about their vulvas again," a reality that is construed as getting "back" to normal. This static future runs counter to the multiple modes of embodied temporality experienced by Sobchack (2010) after she underwent a leg amputation:

> Although my sense of my bodily morphology was newly incorporated in form and thus inaugurated "new sites of projection and identification and new bodily possibilities," it needs emphasis that it was also familiar in function. Thus, phenomenologically, it is not a contradiction to say that my corporeal figuration was, at once, *both new and renewed*. (63; Sobchack is quoting Weiss 1999, 37; my emphasis.)

This temporal complexity has also been addressed by a number of queer theorists (Edelman 2004; Freeman 2005; Halberstam 2005) who call "normative temporal regimes" (Freeman 2007, 165) of procreative heteronormativity into question and who imagine "other anticipations, even other modes of anticipating" (166) experience. In this model, regressive or nonlinear modes of being can constitute detachment and liberatory potential rather than forms of backwardness; they are experiential genres that emphasize the freedom with which people like Daphne can imagine their bodies outside of a linear past where pain and no-pain are mutually and temporally exclusive. These are temporalities that reimagine "queer" as a set of possibilities produced through historical and embodied difference and through which the manipulation of time is viewed as "a way to produce both bodies and relationalities" (159).

Neuropsychologists suggest that body images are contoured by these imaginaries, particularly by the practices they subtend. "Behavioral, emotional and cognitive relevance must cohere," argue Berlucchi and Aglioti (1997, 563), in order for "an integrated awareness [of the body] to develop"; these integrated schemas are gradually refined through "systematic interactions between tactile, proprioceptive and vestibular inputs" (560). Physical therapists like Cathy and Hanna depart from this set of propositions and invite their patients to locate multiple pathways through which they might make their genitals and pelvic floor *matter*—to themselves as well as their partners. This approach "drives the experience of body-ownership" (De Preester and Tsakiris 2009, 313), providing patients with (sexual) bodies increasingly able to accept, control, "dialogue with" (see chapter 4), or redirect their pain sensations. With the deliberate deployment of multisensory and overlapping therapeutic techniques (e.g., biofeedback, therapeutic touching, verbal affirmations, partner participation), these therapists attempt to ensure that the overdetermined pain conditions of vulvodynia and VVS are understood within an equally complex resolution pathway. In undoing a patient's "holding pattern," for example, Cathy delineates the *numerous* channels through which these states of hypervigilance have been developed; this primes her patients for continued relationships with their individually configured experiences of pain, as well as with the emotional, physical, and social experiences through which these patterns have become embodied (De Preester and Tsakiris 2009; Weiss 1999).

During one of Libby's sessions, Cathy and I stepped out for several minutes so that David and Libby could, in Cathy's words, "play" with the biofeedback monitor in the context of their own intimacy. As we gingerly stepped back into the room, Cathy whispered to me how delighted she was that they were doing this as a couple before asking them, "How's it going?":

DAVID: She's been at a 0.8. [On a scale of 0.0 to 5.0; 1.0 and under was the goal.]
CATHY: Can you feel when she goes there?
DAVID: I think so.
CATHY: Do you feel it Libby?
LIBBY: Not really.
CATHY: That's the tricky part for you. The win today is that you were able to get that down to below a 1.0. What does that feel like? Do I experience it in my heart, [my] head, my vagina, where?

Here, Cathy proposes a "morphological fantasy" (Butler 1993; Weiss 1999) for Libby to inhabit. Using a combination of tools—partner, verbal exchange, and Libby's emerging bodily awareness—Cathy redistributes Libby's pelvic floor, allowing it to reside in her "heart, head, . . . vagina, [or] where?" Without asking Libby to explicitly name anything, Cathy nonetheless offers her an opportunity to develop a dynamic and evolving relationship with her pelvic floor, one based neither in *alienation* nor disavowal. Although it is unclear to neuropsychologists what percentage of our body images are innate (e.g., how much of a vulva is "available" at birth), empirical evidence strongly suggests that these schemas are highly plastic and adaptive to the specifics of an individual's experiential environment (De Preester and Tsakiris 2009; Fisher and Cleveland 1968; Kurtz and Prestera 1976; Knoblich et al. 2006; Schilder 1950). The question with which I am engaging, then, is whether Cathy is helping Libby to *re*collect a pelvic floor that had *once* been (unproblematically) hers, as Merleau-Ponty would have us believe, or, in (re)locating her vulva to places within her "heart, head, [and] vagina," is Libby mapping out new genital and bodily terrain (Mol 2002; Sobchack 2010)?

This question points to important work that remains to be done at the nexus of feminism, sexuality studies, and neuroscience. Are patients like Libby and Daphne acquiring "new" vulvas or recovering "old" ones (Salamon 2010)? Can bodies or body parts be "lost" through cultural erasure? Are genitalia, particularly nonprocreative genitalia, part of the "pre-existing body model" (De Preester and Tsakiris 2009, 307) that at least partially determines the body's ability to accommodate new material possibilities? Do redistributed vulvas constitute reconfigured or extended body schemas? De Preester and Tsakiris (2009) argue that the body image can incorporate new realities but only if an existing model contains "substantial prior information about the body" (315); they refer to this deeply internal mode of (re)organization as "body-ownership" and suggest that "the fit between current stimulation and [a preexisting body] model can ... provide a criterion for distinguishing between ownership and disownership" (315) of bodily realities. Following this, cultural critics Cartwright and Goldfarb (2006) use the language of interpellation, suggesting that the body must ultimately be "*hailed* by" (130) new material realities (such as a robust or juicy vulva) in order for the body to be "radically and deeply reorganized" (2006, 130). Drawing on ongoing personal experience, critical theorist and amputee Vivian Sobchack (2010) offers a phenomenological account of this process, describing the body work in which she routinely and necessarily engages in order to accommodate both her prosthesis and the sensations of her phantom limb. Infused with the "new bodily possibilities" (Weiss 1999, 37) made available by her amputation, Sobchack describes a body that actively participates in its reorganization. Through a process she calls "grasping," Sobchack suggests that her phantom/stump and her prosthesis "occupied, thickened and substantiated [one another] so that they made sense to me and became corporeally integrated and lived as my own body" (2010, 63).

I suggest that body model *in*stability, or what Cartwright and Goldfarb (2006) refer to as a "mismatch between material body and body image" (135), dominates the lived genital experiences of many women. In the contemporary United States, the "postural, anatomical and visual clues" (De Preester and Tsakiris 2009, 315) about vulvas contribute to their inconsequence rather than their integrity. Without the sustained efforts of clinicians and therapists—verbal and tactile encouragement, anatomical instruction and clarification, and ongoing emotional support—the women with whom I worked were challenged to "grasp" their vulvas in ways that could promote body ownership. And while I do not wish to suggest that Sobchack's own recuperative process was not also fraught with cultural cues and expectations (e.g., regarding "whole" bodies), I do wish to draw a distinction between body parts "lost" through disease or injury and those lost through attrition. If, like arms and legs, genitals are neurologically "available" at birth, it is important to recognize the number and variety of ways that this availability is compromised in the case of vulvar stability. In other words, though PT may offer vulvar pain patients a "new and renewed" genital body image, as prosthetic work did for Sobchack, the question of how novel these body parts might actually feel merits further exploration. Body parts, like vulvas, that acquire an absence over the course of a lifetime may require other efforts or treatment modalities in order to be reorganized and owned.

But regardless of the vulva's relative presence on or in a woman's innate corporeal map, I contend that there is nothing natural about the states of alienation and phallic/reproductive orientation that she collects without incident in her lifetime. The women described in this book have accumulated the effects of vulvar inconsequence, whether or not their genitalia are a perceptual component of inherited, likely crude, and "probably normative" (De Preester and Tsakiris 2009, 315) postural schemas (Berlucchi and Aglioti 1997; Knoblich et al. 2006). Whether Judy, Joan, and Mary Hudson needed to *return* to an imagined corporeal past or to develop *novel* and future-directed vulvar schemas in order to notice the slow-but-steady erosion of their external genitalia is less important to me than why and how *neither* of these options were available throughout these quite poignant material erasures.

The approaches taken by therapists like Cathy, Lisa, Hanna, and Joy provide patients with access to a body image that contains an uncontaminated and germane vulva, and they do so with language that allows patients to choose the paths most meaningful to them. Indeed, Cathy's techniques, particularly the verbal affirmations that accompany almost all of her bodywork, do not index any clear temporal trajectory through which her patients might (re)locate their genital selves, other than the one in the immediate present of the work they are doing together:

> CATHY (while doing "internal work" on Libby): I'm comfortable slowing down. I'm comfortable relaxing. I'm learning to be comfortable opening and relaxing the pelvic floor. I'm in control. I'm learning to be comfortable.

During this kind of work, patients can imagine vulvas and pelvic floors in less pain, at another time and in another place. Libby, to whom we will return momentarily, came to develop a body image that eventually included the touch of her husband, David, as well as ongoing conversations and feelings about her body, pleasure, and the sexual behaviors in which they were learning to engage. In a morphology that was not dependent on a pain-free past or a future, the experience of possibility could remain constant and Libby could remain capable of being "surprised" by her own body (Blackman 2010, 5).

GENERATIVE POWER

During my fieldwork, I witnessed the possibilities and novel realities produced and sustained by patients, physicians, and physical therapists, all of whom were working to help patients "grasp" their pain. I also observed an expanding array of discursive and material apparatuses that facilitated the circulation of vulvodynia and VVS beyond the clinic and the bodies of symptomatic women. These included the burgeoning expertise of former residents who were establishing practices of their own, transporting their vulvar knowledge to other institutions, as well as the growing economic success of sex therapist Jill's brother-in-law, who developed and supplied therapeutic dilators to both the clinic and the local, woman-owned sex boutique (called It's My Pleasure) where some patients chose to purchase them. Indeed, it was It's My Pleasure's sale of these dilators that helped to crystallize my understanding of the role that material products played in the cultural realization of vulvar pain. As the store's staff grew increasingly savvy about the use of non-pleasure-oriented phallic products for women unable to engage in heteronormative intercourse, and VHC patients came to know more about the queer and feminist sexual practices embraced by the store, I came to think more about the possible—and unpredictably proliferative—effects of these pain conditions.

It is useful to turn to Foucault, specifically his concept of productive power, to think a bit more about the incitement-filled discourses and practices that proliferate around the experience of vulvar pain. In *The History of Sexuality*, Foucault argues that modern, or productive, power is pernicious rather than overtly repressive, primarily because "it is produced from one moment to the next, at every point, or rather in every relation from one point to another. Power is everywhere not because it embraces everything, but because it comes from everywhere" (1990a, 93). In the same text, Foucault also describes a modern "deployment of sexuality," a regime that has its "reason for being, not in reproducing itself, but in proliferating, innovating, annexing, creating, and penetrating bodies in an increasingly detailed way" (107). This "deployment," he continues, is linked to the economy through the physical sensations of a "body that produces and consumes" (107) rather than a sovereign rule of law. Through this analytic framework, we can readily see the work of vulvar pain researchers, the lobbying efforts of the NVA, diagnostic codes, and the taxonomic strategies of clinicians striving to elaborate a more precise nomenclature for these disease conditions as separate-yet-interlocking elements of a "biopower" (140) that is produced and (re)produces itself with an increasingly detailed set of vulvar knowledges.

Much of what is *generated* by vulvar pain, then, is consistent with a biopower that, though capable of producing the tactics by which it can be resisted, is structured through a system of meanings and techniques through which the female sexual body garners the scrupulous and disciplinary attentions of scientific medicine (Balsamo 1999; Bush 2000; Terry 1989).

Thus far, I have attempted to demonstrate that the increasingly technical supervision of *dis-eased* female genitalia reflects neither a benevolent and progressive vulvar incorporation (into the culture at large), nor an uncomplicated and patriarchal (re)appropriation of the female sexual body. Rather, actors and institutions layer themselves around and within competing and converging ideologies about the proper "place" of bodies that are heterosexually and reproductively noncompliant. Within these dynamic social fields, there is room for the resistance that Foucault argues is always coterminous with (bio)power, as well as for strategies and occurrences that exceed these circuits of power and resistance. Based in PT and in the bodily imaginaries cultivated by (some) patients, these latter behaviors and beliefs index a something *else* to which an accepted and resolving vulvar disease condition can point: alternative (hetero)sexualities, vulvar-inclusive morphologies, and an awareness of nonhierarchical bodily difference.

Foucault describes the discursive apparatuses through which we develop our sexual subjectivity, the "sexuality" historicized in his text. He contrasts these discourses with a realm of "bodies and pleasures" that are neither circumscribed by nor incorporated within institutional networks of power:

> It is the agency of sex that we must break away from, if we aim—through a tactical reversal of the various mechanisms of sexuality—to counter the grips of power with the claims of bodies, pleasures, and knowledges, in their multiplicity and their possibility of resistance. The rallying point for the counterattack against the deployment of sexuality ought not to be sex-desire, but bodies and pleasures. (1990a, 157)

Noteworthy here are the "claims" made by Foucault's "bodies and pleasures," which, conceptualized with the same multiple, resistant, and possible terms through which he defines power, seem to undermine his contrast between (natural?) bodily proclivities and those structured by the "complex machinery" (68) of (bio)power. I want to use this piece of Foucault's argument as a framework through which to interpret the recuperating bodies to which I now turn, bodies that I read as both generative and full of possibility, but that I do not see operating *outside of* the diffuse discourses that many of us have come to call "sexuality" (Fausto-Sterling 2000). Rather, in coming to inhabit novel *body images*, Daphne and her peers may demonstrate no more—nor less—than what some corporeal feminists refer to as "modalities of becoming" (Colebrook 2000, 89)—"bodily styles, habits, [and] practices, whose logic entails that one preference, one modality excludes or makes difficult other possibilities" (Grosz 1994, 191).

Unlike Foucault, my aim is not to replace a "repressive hypothesis" (1990a, 10) with a distinct set of ideas about productivity and proliferation. Rather, I am suggesting that women (with vulvar pain) can invest in corporeal strategies that access pasts, presents, and futures as clean as they are contaminated, in which their experience is not temporally or experientially compartmentalized. Daphne, for example, may well have had a greater capacity to "hold her ground" as a young girl, before she had accumulated the *unwanted genital experience* that came to inform her VVS, but this doesn't mean that her hypothetically less sullied body image ought to be the object of her recovery. Daphne's body has "become-meaningful" (Colebrook 2000) to her through an enormous and idiosyncratic range of discourses, behaviors, and affects, and the plastic nature of her body schema can adapt to nostalgic as well as emergent orientations. The body, in other words, with which she can have painless vaginal intercourse is an amalgam of imaginaries that she negotiates, like Foucault's notion of modern power, "from one moment to the next" (1990a, 93).

Control (at a Whole Other Level)

Like the VHC doctors and physical therapists, Braidotti (2002) sees a lot of work ahead for women with vulvar pain, specifically those who "yearn for change": They "cannot shed their old skin like snakes. This kind of in-depth change requires instead great care and attention" (26). Braidotti's bodily imaginaries are not spatially or temporally limited. In other words, there is no one place or time through which a woman can or should locate her (sexual) subjectivity. Rather,

> "becoming" is about repetition, . . . affinities and the capacity both to sustain and generate interconnectedness. Flows of connection . . . mark processes of communication and mutual contamination of states of experience. As such, the steps of "becoming" are neither reproduction nor imitation, but rather empathic proximity and intensive interconnectedness. It is impossible to render these processes in the language of linearity. (8)

As female bodies prevented from accessing the phallocentric sexual order, Daphne and her peers live with "memories of the non-dominant kind" (Braidotti 2002, 8), memories that can function as both sites *for* production and spaces *of* return. Women with and without vulvar pain will continue to accumulate the "discursive practices, imaginary identifications [and] ideological beliefs" (Braidotti 2002, 26) through which their sexualities are simultaneously constituted as alien, absent, and *too much* to deal with. Yet through a sustained and material engagement with their pain-filled vulvas, that is, the work of PT and a reconfigured body image, symptomatic women have an opportunity to interact with these social phenomena in novel and unpredictably productive ways.

Such strategies are crucial for diagnosed women who are (re)encountering partners and husbands from whom they have long been physically and sometimes emotionally separated. The sexual intercourse that patients begin to imagine (and eventually have) requires negotiations with not only their pasts and futures, but with body images that include newly informed genitalia. Women who, with or without clinical encouragement, are invested in their pain being *absent* after surgery, lidocaine, or laser therapy, may not develop the resources with which to confront the "intensive interconnectedness" that physical intimacy can involve, particularly when it includes the "common negative experiences of . . . embarrassment, disappointment and boredom" (Cameron and Kulick 2003, iv). When Jessica, another patient who "got better" during the postsurgical work she did with Cathy, noticed that she sometimes did not want to be sexual when her husband did, she realized that she had never cultivated the skills to refuse him for any reason other than pain. In its absence, she struggled to balance a tenuous mixture of desire, guilt, fear (e.g., of a recurrence), and a newly emerging sexual subjectivity—one that was as informed by the intimacy through which she and her husband had survived approximately five sexless years, as it was by her emerging corporeal autonomy. Jessica had not, therefore, *returned* to a normal marriage; rather, she related to a situation through which she more closely resembled women *without* vulvar pain (who are also in marriages with occasionally asymmetric needs and desires) with and through a body image yet to produce its own version of sexual refusal.

I want to now take an extended turn toward Libby and David, in order to more fully elaborate the two levels of tension—temporal and interpersonal—negotiated by patients whose pelvic floors and genitalia are fleshed out and "become-meaningful." During our interview, which took place about two months before her surgery and one month before she began PT, Libby told me that she was excited about sex therapy because she was "looking forward to learning what to do."[2] She and David were both virgins when they married, and her pain had begun less than three months later; subsequently, neither of them felt terribly confident that they knew what they were doing regarding the physically intimate aspects of their relationship. This facet of Libby's experience intrigued me, as it was only marginally related to her pain condition; but as she opened up to me about this issue, I wondered about whether she and David would have pursued sex therapy if her pain hadn't developed, or whether they, like lots of other couples, would have just figured out "what to do." This is a crucial issue for us to consider, in that it not only indexes the numerous other women about whom this book is written (who might feel like Libby) but also speaks to the queer nature of vulvar pain conditions.

"In everyday life," argues Stevi Jackson (2006), "normative heterosexuality is available as a resource, but is also an ongoing accomplishment[:] we 'do' gender, sexuality and heterosexuality . . . through everyday interpretive interaction . . . [and] actual practical activities [like] having sex" (114). But for David and Libby, vulvar pain had disrupted this heteronormative narrative, making it possible for them to acquire an alternative set of sexual behaviors, as well as to develop a relationship with penetrative intercourse that was decidedly "unnatural." Libby resembled the majority of my informants in that she and David did not use the four years between her symptom onset and diagnosis to explore, even moderately, "other ways" of being sexual. She was unique, however, in her frank acknowledgment of the questions and obstacles she anticipated in incorporating this allegedly natural behavior (back) into their intimate life. What I hope to show here is that though Libby and David did learn how to have penetrative intercourse in their sessions with Cathy, these lessons did not constitute a simple transmission of heterosexuality in its hegemonic form. Rather, in learning about vaginal-penile intercourse via Libby's vulva and pelvic floor, David and Libby acquired a mode of physical intimacy that "question[ed the] . . . stranglehold and naturalized status" (Beasley 2010, 208) of the phallocentric, compartmentalized, and teleological sexual discourses through which they may have otherwise come to know each other's genital bodies.

Without a routine experience of vaginal penetration, Libby first had to make anatomical and affective room for both David and her therapeutic dilators. Her strategy, one that I heard from other informants, was to make the dilators "medical" and to make David "sexual":

CATHY: Whatcha been doing [with] your program at home?
LIBBY: [My period] killed everything for a week, [but] we've used the dilator three times. And David's finger one time.

CATHY: [?]

LIBBY: I'm trying very hard to make the dilator a medical/sexual. If I put David in the category of my PT, and my healing [?], then he's still there later. Does that make sense? [The dilators are a medical] thing and David [is] a sexual thing. I'm not sure where on the spectrum to put all this stuff, so I decided that the dilator is medical and David is . . . [She looks at David.]

CATHY: Yeah, good discussions to have. . . . So, given the view that you're taking, how did you use the dilator? Yourself or with David?

LIBBY: With David, but it was as a medical/PT thing. If we did anything else, it wasn't the result of the dilator, it was the result of spending time together.

CATHY: Just so you have the information, some people use the dilators as sexual play.

LIBBY: Yeah, but if it's part of PT . . .

CATHY: It would be very good to see Karen [the therapist] now. She could step in here, and give you some clarity.

[There is some logistical negotiation about this for a few minutes.]

LIBBY: [Moves toward David and pats him on the leg.] We can't have sex until I've used all three dilators. Then it's your turn. [My field notes say, "very cute."]

CATHY: *Yeah!* And if while you're fooling around, he wants to slip a finger in . . . Allow yourself to feel aroused. *Allow* yourself to go to those places.

We can see that what Cathy seems to want in this exchange is for Libby to integrate her penetrative experiences, allowing a dilator and a husband to occupy the same discursive space in Libby's shifting sexual body image. But the key term here is "shifting," in that Libby—by *becoming* a body that can tolerate and imagine penetration—is experimenting with corporeal strategies specific to her own experience, "destabilizing," in Beasley's (2010) words, the "reductive assumptions about the political possibilities" (208) of heterosexuality.

In short, what Cathy perceives as compartmental thinking may instead reflect a vaginal body image that is multiple, one that exceeds and alters the extant and hegemonic genital horizon through which Cathy interprets Libby's sexuality (Irigaray 1993b; Mol 2002). "This is where close attention to what actually goes on in the everyday dimension of social practice can help," argues Jackson (2006, 114), since norms are visible "not only in their effects, but in the act of reproducing them" (114). Because David and Libby are disrupting, rather than reproducing, heterosexual norms, even Cathy lapses into a narrative that consolidates Libby's vaginal experience into a singular one, rather than allowing it to proliferate into the multiple and intersubjective bodies that many argue are central to embodied experience (Blackman 2010; Csordas 1994; Mol 2002; Weiss 1999). Cathy's efforts to contain Libby's radical bodily imaginary allows us to see the process of heteronormalization in action, as well as how it intersects with other normative modes of being, such as linear temporality and unitary embodiment.

Oliver Sacks (1987, 1995) has written extensively about the experiential worlds of neurologically impaired individuals, making a compelling case for the generative nature of these injuries. Mixing the personal narratives of his patients with those constructed by neurological medicine, Sacks argues that "defects, disorders, [and] diseases . . . can play a paradoxical role, by bringing out latent powers, developments, evolutions, forms of life, that might never be seen, or even be imaginable, in their absence" (1995, xvi). He continues: "While one may be horrified by the ravages of developmental disorder or disease, one may sometimes see them as creative too—for if they destroy particular paths, particular ways of doing things, they may force the nervous system into making other paths and ways, force on it an unexpected growth and evolution" (xvi).

I want to reiterate that the bodies that might benefit from this neurological creativity are as cultural and collective as they are individually physiological. In his ethnography of the virtual world Second Life (SL), Tom Boellstorff (2010) addresses these issues in his discussion of the alternative avatars, or "alts," employed by some SL users. Alts are created by users not only to play the role of another gender or bodily capacity; they also allow a person in "actual life" (21) to inhabit multiple bodies simultaneously, reconfiguring what users might regard as a self or person. And though the question of whether Libby's various pelvic floors constitute distinct selves or multiple dimensions of one self is beyond the scope of the present discussion, we can use the figure of the "alt" to extend Sacks's proposition. That is, in generating multiple vaginas, around which multiple perspectives regarding her pain and her sexual relationship can cohere, Libby demonstrates that imagining another (form of a) body can have material correlates.

In the following exchange between Libby, David, and Cathy, we once again witness the eruptive potential of her pain:

> CATHY: Do you want David's touch again?
> LIBBY: I'm okay.
> CATHY: I'm going to challenge your "I'm okay." One thing to think about—'cause this can help with sex and intercourse too—think about what feels good: "I'm okay," but "How could I be better?" [David has moved so that he is standing behind Libby's head and is stroking her hair. Libby is on her back on the exam table and Cathy is working "internally," in Libby's vagina. Cathy says she "felt [an] area" loosen as David repositioned himself.]
> CATHY: You'll want to think about this as you take [this work] into the bedroom; to be able to verbalize, because it's not easy for a lot of women to do that. [To David:] [You must be able to] **take it** as she learns to tell you what doesn't feel good . . . There is a difference between "taking it" with resistance and "taking it" openly. [To Libby:] He has his own work to do around that—to take what you say and truly, truly accept it. [After noticing that Libby has been glancing toward David:] She keeps looking at you, David. [Then, to Libby:] What are you thinking?
> LIBBY: I want to know what he is thinking.
> CATHY: Would you like to ask him?

DAVID: [I don't think I have a problem, but I can] look at it. [Cathy is trying to help David ask Libby something directly, but they are evasive. There is some stumbling around.]
CATHY: *If you leave here and something comes up—it can be **very hard** to tell him, and that can be important, because he may not know.*
LIBBY: [I think we've] *always been able to talk about it. It's the strength of our relationship.*
CATHY: *Okay, you tightened up [when you said] that.* (Italic and bold emphases mine; see further discussion.)

Prior to her work with Cathy, Libby had discerned that there were things about sex that she (and David) still wanted to learn. A more conventional physical therapist may have worked effectively with Libby's pelvic floor but made no explicit connections with the rest of her body, her relationship, or "the way that [she *did*] life." The practices of sex therapists vary even more widely, and, had they pursued it, Libby and David may have encountered anyone from a conservative Christian invested in gendered hierarchies (they were both practicing Seventh-day Adventists) to someone like Jill, who was both more liberal and more knowledgeable about vulvar pain. At any point with those therapists, however, it is doubtful that Libby's pelvic floor would have communicated what it did in the previous exchange. That is, by deploying a set of "systematic interactions between [a variety of] inputs" (Berlucchi and Aglioti 1997, 560), Cathy guided Libby through a morphological shift in her body image and gently established its salience in the physical intimacy that Libby and David were attempting to (re)establish.

Imagine That: Corporeal Feminism and Anthropology

I want to address the last three utterances of the previous dialogue by considering the likelihood of them pertaining to *any* heterosexual couple in the contemporary United States, regardless of their relationship to vulvar pain. Based on the narratives that I've collected in twenty-five years of talking with women about sex, I would venture to say that it is considerable. In this imagined group of heteronormative individuals, some might find their way to sex therapists who, though unable to deliver such convincing bodily evidence, would typically make similar assertions about the importance of good communication. Conversely, clinicians capable of detecting *pelvic floor myalgia* would locate relief measures in the individual—and physiological—bodies of the patients under their care. In this second scenario particularly, the burden of dis-ease is assumed by women, making them the identified patients of a syndrome that I argue is culturally endemic. Approaches to (hetero)sexual distress that intervene at the level of the individual may be an improvement over a Freudian repudiation of the "real" traumatic stories of girls and women, as well as over the psychosomatic gloss that vulvar pain received for at least a decade. But without a broader analytic stance—one that bridges a collective female structure of feeling with (some of) its more individual bodily expressions—we can neither access the proliferative nature of (political) power nor hope to deploy it productively.

Twenty-eight years ago, in their seminal prolegomenon, Scheper-Hughes and Lock (1987) persuaded critical medical anthropologists to think about the body at three levels: individual, social, and political. I invoke these famous bodies once again in order to draw explicit attention to the political nature of shifting and multiple body images and of the discursive redistribution of vulvar pain. That is, if genitals are to be taken into account by the feminist project of elaborating nonhierarchical sexual difference, then they must be adequately differentiated *across* those bodies that position themselves as women (Fausto-Sterling 2000; Karkazis 2008; Salamon 2010). Rather than as indexes of phallic/reproductive capacity, sexual orientation, or dichotomized physical normality, vaginas and vulvas can "sex" bodies female only inasmuch as they correspond to bodily sensations, experiences, and intersubjective situations through which "female" or "woman" is also established

(Fisher 2011). In other words, vulvas in themselves do not render bodies—or people—female. Any role they *do* play in our understanding of gender's materiality, however dynamic, must account for their widest diversity. Colebrook (2000) argues that a conventionally feminist "appeal to equality assumes that gender differences are imposed on otherwise equal beings, . . . thereby preclud[ing] the possibility that different types of bodies might demand different forms of political recognition" (76). Through theories of corporeal *difference*, however, bodies with genitals that have been cut, eroded, surgically affirmed, mechanically altered, or that simply don't "work"—and yet are structurally sexed as female—can be more sensibly interpreted as generative and proliferative rather than as lacking or defective.

Allowing Libby to have both "medical" and "sexual" pelvic floors, then, is as much of a political project as it is individual or interpersonal, as it allows for the affirmative, multiple, and *bodily* difference through which many feminists hope to redistribute (political) power. Similarly, in aligning women with *and without* vulvar pain around potentially similar struggles to communicate *from* their genital bodies, we find even more— and unpredictable—sites of articulation from which to build a feminist corporeal politics. Unsurprisingly, the key organizing device here is perspective. Adjusting the analytical focus of our questions about these symptoms, from one that is person- and defect-centered to one that is multiple and generative, makes it increasingly possible to locate the numerous nodes around which a corporeal-genital politics could emerge. For once we begin *listening* to the narratives of symptomatic women, we find that vulvodynia and VVS are less a (physiological) problem for women with "pretty white skin" than a set of avenues through which women can choose to "exit from the universal mode defined by man, toward a radical version of heterosexuality" (Braidotti 2002, 27).

Such exit strategies abound once we begin to view genital pain in this way. For example, although *dyspareunia* is a common clinical term used to denote "pain with intercourse," it is rarely used in the management of vulvar pain and commonly used to describe pain associated with deeper thrusting. But this seemingly reasonable and benign distinction enacts a separation between women diagnosed with chronic vulvar pain and those diagnosed with chronic *pelvic* pain (CPP) (the main symptom of which is *dyspareunia*), many of whom are African American. By employing—indeed constructing—a new language with which to describe vulvar-based conditions (*allodynia*, *hyperalgesia*), clinicians distinctly code these women, thwarting their respective abilities to identify and align themselves around their common experience of sexually prohibitive pain. Similarly, transexual and transgender individuals whose vulvas may anatomically vary from a constructed norm might, via discourses of multiplicity, experience an affective and corporeal camaraderie with *LS* or *LP* patients, whose labia also appear at times to be "nonnormative." Again, such alliances might not only produce new forms of political and social collectivity. They can also redefine the female genital imaginaries of women in the United States, from those defined through inconsequence and disparagement to novel discourses and practices characterized by difference and possibility (Carrillo Rowe 2008).

With greater collective access to *these* imaginaries. Libby need not remain the only woman whose vulvar *body image* is redistributed to her "heart, head, [or] vagina." Indeed, limiting the meaning of Libby's work to her personal relationship with David actively distances her from the radical heterosexuality called for by so many feminists and potentiated by the bodily experiences of my informants. In attending to these women as an anthropologist and disseminating my own interpretations of their corporeal transformations, I hope to open their experience to the legions of other (US) women who might benefit from an "intentional and active ensemble of . . . corporeal alignment and movement" (Sobchack 2010, 62) toward their vulvas. Living with the genital dis-ease described in these pages, many of these women suffer their sexual dissatisfaction and disappointments in relatively silenced, resigned, and medically pathologized terms (Cacchioni and Wolkowitz 2011). In distributing genital affect across the body, away from the sites of contamination that reinforce this silence, more women may find it easier to speak their experience with greater clarity, increasingly able to "hold [their] ground" while they unsettle the territory upon which phallocentric heterosexuality is currently lived.

GIFTS

This Is Like a Broken Bone I Didn't Know About. I'm Going to Get a Cast.

> CHRISTINE: Do you think that your symptoms have given you anything?
> CLAIR: Um, no. I would say I feel deprived, all the time. Always deprived.
> —From my field notes

David was not the only male partner I met whose ideas about masculine sexuality were both tested and revised by vulvar pain. Jim, the budding mortician mentioned in chapter 5, who also typically accompanied Brigette to her appointments, was explicit about this during our interview. I had asked Brigette whether she felt like her symptoms had "given" her anything, whether they had provided her with anything to which she hadn't previously had access. After she briefly described the "better understanding of people who . . . have other problems in that area," she indicated that Jim could also answer this question, which he did:

> The word *humbling* comes to my mind. . . . Just having to be more sensitive to the situation, for myself, emotionally, to support her in that. It's something I need to remind myself of, of why, and not just "We're here," and not just pull the general male "get upset something's wrong and not going our way" thing.

As we continued talking, I asked each of them if they believed that their sexual relationship had been challenged by vulvar pain. Brigette responded that Jim was now "a lot more understanding than even when we started dating. He cares about me and doesn't want to put me in a position of pain," to which Jim added that he had experienced "more growth emotionally" and had more (sexual) "patience."

> CHRISTINE: Do you two think that you have anything to teach other couples, whether they have this problem or not?
> BRIGETTE: Just if someone's not able to be open and talk about it, then you shouldn't be with them.
> CHRISTINE: Did you think that before?
> BRIGETTE: No, I was kind of like, "Keep it quiet." But now, after dating Jim, everything is so much easier. I didn't even realize it because I'd never had it, but it's so much easier. I can't keep it quiet with him.

I want to turn now to the voices of several of my informants who, like Brigette and Jim, answered my questions about what the experience of vulvar pain had given them. My informants and I would not have had these exchanges had I not posed the questions. But I wanted to draw them out about this topic in order to find out if their embodied experiences reflected, resonated with, or contradicted the (broader) social body that both structured and was structured by their individual ones (Bourdieu 1977, 1984; Douglas 1966, 1973; Martin 1994, 2009; Scheper-Hughes and Lock 1987). How, for example, did Brigette's newfound inability to "keep it quiet" correlate with broader social processes that governed her desire to speak? Given the generation of both affects *and* effects from the discursive and institutional sides of vulvar pain, I was curious about its more internally felt striations.

In response to my questions, women were both insightful and forthcoming, voicing some overarching themes that I describe later. I position these themes through Sacks's discussion of disease-related generativity—that is, as manifestations of "'a whole new world,' which the rest of us, distracted by [normality], are insensitive to[;] . . . a new state of sensibility and being" (1995, 38–39) to which symptomatic women have greater access and awareness. A theme that surfaced both quickly and consistently is the one to which Isabelle gives such eloquent voice at the start of this chapter: the bodily confidence and awareness that patients believe they will carry forever. Often spoken in words similar to Isabelle's, women spoke of the gratification they felt toward their newly acquired vocabulary and anatomical knowledge, and the greater sense of bodily mastery that went along with them. Mya, for example, struggled to bring her body across the intimate threshold required by PT but experienced, nonetheless, a developmental shift in learning how to describe her symptoms to a provider. In telling me about how she planned to teach her daughter the "correct" words, Mya explained that this decision was a direct result of her experiences with gynecological medicine:

> MYA: I want her, when she goes to the doctor . . . That was *really* hard for me to say.
> CHRISTINE: [Can you tell] me about that?
> MYA: You know, people don't realize that [laughs]. . . . Um, well, you know, because of how I grew up, no one . . . *talked* about it. Even when I would go to my friend, you know, you could, you could . . . get around not saying any words. And, and so that's how I've always been, and so it wasn't until really that, like, the doctors aren't getting it. I need to . . . , you know, des—, describe. To say what *part* . . . of my vagina it is and . . .

On my prompt, Mya admitted that she had been able to point and gesture toward the areas of her genitals that were most painful, but that even this had come with considerable awkwardness. When I asked Mya about the word "vulva," after she told me that she would teach her daughter "vagina," she responded, "Yeah, that's a new one for me, too." Again on my prompt, Mya indicated that "vulva" felt "a little dirtier," but she stressed nevertheless that her new vocabulary made her feel like she was "taken more seriously," instead of like the "little girl in their world" whose pain had been effectively shelved for close to a decade.

Although Ashley was initially somewhat tentative, she also indicated that a closer encounter with her vulva hadn't been *all* bad:

> CHRISTINE: What do you think that these symptoms have given you?
> ASHLEY: Definitely a better understanding of my anatomy.
> CHRISTINE: Are you grateful for that? Is that a good thing?
> ASHLEY: Possibly. I hope that, um, it doesn't have to be just a way of, you know, having learned that, . . . that I . . . have some opportunity to use it, other than just the pain management issue. Um, so . . . [it] probably would have been better to learn that earlier [laughs], would have been a little more . . . .

Ashley's response nicely captures the tinges of ambivalent regret that often coexisted with this new knowledge, the wincing with which women confronted the ignorance and lack of interest that preceded their burgeoning vulvar expertise. Isabelle reiterated, "I deal with it better now than before. I have more information, . . . no more embarrassment. And I guess . . . I was before. [It just] didn't feel natural to have anybody poke around down there." And for some, like Brigette, this emerging sense of bodily mastery played out in the sexual "holding their ground" of which they now felt more capable.

> CHRISTINE: Do you think you would ever have sex in pain again?
> BRIGETTE: No, if it's really bad, no.

Finally, Tina, who, at the time that I interviewed her, was still working with Dr. Robichaud to get her *lichen planus* under control, spoke about it this way:

> CHRISTINE: What do you think that these symptoms have given you?
> TINA: Well, it's made me a little more aware of my body, you know. *And* that these things *happen*. And that, uh, you know, not to go ten years without going to a doctor, for one. And, like everything else that's gone

wrong with my body, it could, it could have possibly been prevented, you know, had I, had I had more sex.
CHRISTINE: Mmm. What makes you say that?
TINA: Well she [Dr. Robichaud] told me, you know, by using, by *doing it* more, it keeps it more elastic.
CHRISTINE: Yeah. Yeah, but, you couldn't have prevented the LP.
TINA: Couldn't [have] prevented it? Well, I would have jumped on it sooner.

Tina went on to say that she hoped her daughter would be able to avoid her mistakes, "that, uh, I hope, hopefully that knowledge gained is something that I can pass on to my daughter." Many of the women that I formally interviewed spoke about this aspect of their symptoms, particularly as they tried to think through why and how "genital education" had been missing from their own lives. Indeed, enough women spoke about the hypothetical talks that they would have with their daughters that it sometimes seemed as if it was a question I had written (it wasn't). These thoughts would often emerge toward the end of our interviews, when I asked women if there was anything that I'd left out, if there was any part of their experience that they felt hadn't been adequately captured by my questions. Mya, whom I quoted earlier, ended our discussion by reiterating that she would be "as open as I can be" with her (special needs) daughter to "make sure that she can be comfortable. I'll talk to her as much as I can." When I asked her if she was referring to vulvar pain and disease, Mya replied, "I *will* tell her parts of it, what's appropriate."

In thinking about the daughters of these patients, I am struck by one more connotation of the word "generation," one that seems particularly apt in this discussion. The real and hypothetical daughters of these women represent something like a new generation of women, one in whom we can store and kindle our hopes that it will be better; that, in the words of Foucault (1990a), "tomorrow sex will be good" (7). I am leaving off the last word of Foucault's iconic sentence (which is "again"), however, because neither I nor my informants (nor many of

Foucault's feminist critics), I would venture to say, believe that it has ever been particularly "good" for a lot of women. But in lining up these optimism-fueled daughters alongside the vulva clinic's residents and medical students, the NVA and NIH grant recipients, the potential audiences for a book like mine, and the tens of thousands of women whom Harvard's epidemiologists have begun to successfully tap with newly configured survey research, we may indeed be facing not just one but several generations of women who will be increasingly—and stunningly—savvy about their vulvar anatomies. Whether this *generation* will constitute a new genital order of the kind that Irigaray (1985a, 1993b, 2004) has insisted is politically necessary in a *female imaginary*, or a paternally co-opted one through which women can achieve greater sexual "autonomy" through procedures like cosmetic labiaplasty remains to be seen. But if the realities of vulvodynia and VVS are eventually disseminated with little more than an "awareness" campaign (King 2008), marked by public service announcements, an early detection protocol, and a color-coded ribbon (my money's on fuchsia), feminists will be left with considerably more biocultural work to do.

He Won't Like That

Libby's pelvic floor reminded Cathy, Libby, and David that talking about sex may not be "the strength of [*any* (hetero)sexual] relationship." But because poor—or missing—sexual communication is nearly impossible to ignore in the face of searing penetrative pain, "talking about it" remained a poignant issue in the lives of many of my informants. This was especially true for older, sometimes postmenopausal, women who were more likely to be diagnosed with *dysesthetic vulvodynia* or one of the *lichens*. In general, these were women who had been quietly tolerating penetrative intercourse on their husbands' terms for twenty years or more and whose current inability to participate was forcing them both to confront the changes that might be necessary in order for them to be sexual together. In one of her visits with Dr. Robichaud, Anharrad put it this way: "I feel like lately my whole life is all about my vulva. And I have to work to not be resentful of my husband, because he wants sex. And he's only getting it once a week, the poor man." Unlike Jessica, who was unaccustomed to refusing her husband on *any* grounds but her VVS pain, Anharrad was struggling with how to refuse sex for *any* reason, given the dynamic of a thirty-plus-year relationship in which her (sexual) needs had rarely been prioritized.

Anharrad was the first patient whose visit I observed at the vulva clinic, and she wanted to share her story for the sake of other women who were potentially "suffering in silence." She was also taking advantage of the opportunity to explore the shifting and potentially generative nature of her sexually based symptoms. During one of my last clinic visits with Anharrad, she told me that she was beginning to wonder if the trials associated with her lichen planus weren't some kind of "punishment" for having "participated dishonestly" in her sexual relationship with her husband "all these years." Anharrad was quite emotional—indeed, she was near tears—as she expressed this to me in the quiet of the exam room, while we waited for Dr. Robichaud to return with prescriptions.

Anharrad told Dr. Robichaud and me that, though her physiological symptoms were the notable "cause" of the disrupted sex in her marriage, "this [dynamic] has always been present in our relationship." This exchange presaged my interview with Sharon, who recounted the equally difficult set of marital-sexual negotiations in which her LP had landed her. During her first clinic visit, Sharon told us that her symptoms began around the time of her menopause, "the year my periods changed." She said that sex had been "uncomfortable[;] not to the point where I wouldn't do it, but I didn't want to." She noted that her symptoms were penetration-based and that she'd "gotten better now figuring out where the pain is." After a few minutes of exchange during which she and Dr. Robichaud began discussing LP as a possible diagnosis (Sharon described "thinner" and "smaller" labia, as well as areas that "were like a razor blade has cut it"), the subject of dilator therapy was raised. Sharon relayed that she'd been given one by a previous provider, and Dr. Robichaud began to emphasize the role they played in maintaining the "caliber [of] the vaginal vault," particularly for patients who had lost some capacity:

> DR. ROBICHAUD: You use the dilator daily when the tissue is intact. So you can *have sex*; [so you can] continue to be intimate with your partner.
> SHARON: That's definitely one of my goals.

On the surface, Sharon presents a straightforward and self-interested desire to (re)engage in penetrative sex with her husband. An attention to the beginning of her narrative, however, where she suggests that she engaged in sex that was painful—*although she did not want to*—helps to uncover a deeper dynamic that structured Sharon's experience of her symptoms. Because her diagnosis was LP, meaning that there were effective treatment options available, Dr. Robichaud did not spend any time exploring what *else* she and her husband might have been able to do; rather, and partly due to its therapeutic purpose, she quickly directed her discussion to the restoration of regular penetration (i.e., with intercourse and dilators) when Sharon indicated that this was one of her "goals." In our interview, however, Sharon—like Anharrad—told me that her husband was "uninterested" in learning about or participating in modes of sexual activity that might be not only less painful for her but perhaps even more pleasurable.

> She says their relationship has been "deteriorating," getting "farther and farther apart" and is "really odd right now." "He would just like to pretend that everything is hunky-dory." I press about new conversations [but] she is vague. On my prompt, she says that she feels like she "need[s] to get better to, to see if this relationship's going to get back into tune again," but that her husband won't "even engage in a conversation about it," even when she returns from the clinic and tries to explain it. "*The way he is, is I have to fix the problem.*"

He knows about dilators but he "doesn't talk about it."
(Field notes; my emphasis.)

Sharon's having to fix the problem sheds new light on her goal of returning to allegedly uncomplicated penetrative sex, as well as her later question to Dr. Robichaud: "Any chance that this will just go away?" Her awareness that her husband would rather pretend that nothing is happening echoed Mya's description of her boyfriend's disappointment when her long-awaited consultation at the clinic didn't "fix" the problem:

> I tell him everything so that he doesn't think I'm full of it. [He's just like,] "I don't want to hear it!" He was [also] disappointed [by the outcome of her visit:] Yeah, 'cause I think he really thought, "Oh good, she's gonna come back cured, and . . . right away we're gonna have sex. [laughs] I think that's what he thought.

Mya described some "pressure and guilt" about her symptoms, but stated that relationship counseling was "not anything we've ever brought up," a reality expressed by several of the women whose stories I am exploring here. When Dr. Robichaud asked Anharrad about her husband's reaction to the idea of seeing a therapist, she told us, "He's from a different generation. He has never believed in that. He won't like it." She elaborated that her work with Dr. Robichaud was just "treating the symptoms, which is the best you can do at this point," reminding us that she could see the underlying (and less easily "manageable") problem. She reiterated, "If I wasn't in a relationship, this wouldn't be an issue."

Having to *deal with* the unexplored, unvisualized, and perhaps un*know*able aspects of a female sexuality that may be based elsewhere—anywhere—is an overwhelming task, one to which many women understandably would rather not turn on their own. In response to the interview question that complemented the one framing this discussion, in other words, "What have these symptoms *taken away* from you?" Sharon eloquently captures this sense of ambivalence, disappointment, and frustrated wonder:

CHRISTINE: Do you think that these symptoms have taken anything away from you?
SHARON: My sex life. My, um, *intimacy* with my husband. But, and I don't know if this has taken that away, or if it's, a lot of things going on and . . .
CHRISTINE: [Is this] pointing to it, maybe?
SHARON: It's definitely [about] enjoy—, definitely *enjoying* my, any sexual relations, . . . intercourse. You know, we still have oral sex, and . . . but that's . . . seems . . . what's considered? . . . seems to be more about him, so [laughs], and less about me. [laughs]
CHRISTINE: Do you think you'll talk about that with him?
SHARON: I don't know.
CHRISTINE: Is it something that you want? From him?
SHARON: No. It's, you know, it's something you don't want to have to . . . I don't know . . .
CHRISTINE: [Do you engage in more] nonpenetrative sex? [Is it] part of what you do? [Do you] do [that] more?
SHARON: We haven't talked about it. . . . This is just kind of the icing on the cake for my husband and I. It's like a weight, you just keep adding to it. It's like, what's next?[3]

Ashley also struggled to articulate her mixed feelings about what she had learned about her body and what she might therefore be able to ask or *expect* from future (sexual) relationships (she was not in one at the time of our interview). Ashley knew—and emphasized—that men had always "got[ten] off easy—we couldn't talk about it, and so they told us it was [us] and we believed them. But now if you've watched Oprah or read a book . . ." But her ambivalence about what to *do* with the information that Oprah and sexuality literature had made available to her was also explicit: "[You] have to ask for it, [and] it's going to be an unpleasant encounter; you have to ask for it twenty times," something that she, like Sharon, was relatively loath to do.

Because her unremitting pain had compelled her to (re)locate her vulva to the center of her attention (at least intermittently), Ashley began to rethink the stakes of both past and future relationships. Noting that "the sale of vibrators is not going to go down in the near future," Ashley stressed that women (like her) were "challenging" received ideas about their inferiorized place on the sexual hierarchy. She continued:

ASHLEY: And I think that, had I stayed married, I probably wouldn't, I mean, that would have, I would have.
CHRISTINE: [You wouldn't] have worried about this?
ASHLEY: Right, because, every time I did, in my marriage you know, I was told something like, "Relax and enjoy it," or whatever. . . . And I just don't, I mean, maybe because it would have reached a point by this time that . . . where that would have been hard. And until it became an inconvenience for my *husband*, I probably wouldn't have [said anything].

Ashley hopes that a partner's "inconvenience" will no longer be the sole arbiter of her ability to speak up about circumstances not to her liking. Admitting that her low (sexual) self-confidence and her belief that men would find her symptoms unacceptable allowed her to "put up with a borderline abusive partner" for a time, Ashley told me that she has "plain quit that" way of thinking and behaving. In thinking about what will be "available" to her on the dating scene, she realizes that she "might have to change things—like talk about what's uncomfortable."

These are the unnerving conversations around which Anharrad, Sharon, and their respective husbands were constructing variously circuitous paths. At the time of their diagnoses, both women were defining "intimacy" in terms that depended upon their (physical) ability to practice the kind of sexual intercourse their husbands had come to expect. When running smoothly, this pattern allowed them to divert and displace the elements and aspects of their sex life that might have been more directly "about" them. Anharrad and Sharon's symptoms troubled a set of sexual—and, for their husbands, *settled*—issues; the "affective proliferation [and] contraction" of their bodies had "implications" for their husbands (Michael and Rosengarten 2012, 4). Indeed, deciding upon and sticking with a treatment plan that involved therapeutic vaginal intercourse became more volatile once each of them had been able to differently perceive their husbands' investments in *their* (thus far underexplored) pleasure. In a sense, Anharrad and Sharon had to decide what kind of marital futures they wanted their symptoms to help them create. In this exchange between Sharon and Dr. Robichaud that occurred at the end of one of her visits, their dialogue about her proper use of lidocaine makes an apt metaphor for the less physiological aspects of her (and her husband's) dis-ease condition:

> SHARON: [Says she wants to try using a larger dilator with some lidocaine.]
> DR. ROBICHAUD: Let me tell you, the two percent gel is sticky. It will probably get on your partner's penis.
> SHARON: He won't like that.
> DR. ROBICHAUD: No, most of them don't. [She goes on to describe how to apply the lidocaine with a cotton ball. She compliments Sharon again on her correct use of the medication.] I can tell that you [are] put[ting it way] up there, up inside.
> SHARON: It's kind of a hard area to reach.
> DR. ROBICHAUD: *It sure is, but it's the most important.* (My emphasis.)

Negativity

Finally, a number of women answered my questions about generativity in terms that were straightforwardly negative. Like Clair, cited at the start of this section, several interviewees felt no need to look for or identify altruistic or "self-helpish" elements of their experience, elements they simply did not believe were there. Mya, for example, although she eventually elaborated on how things would be different for her daughter, was initially quite clear that her symptoms had not given her "anything but grief": "You know, it's probably made me, like if, if someone was ever gonna talk to me . . . ? Then I could probably be 'Oh yeah, I understand what you're saying.' So, as far as like maybe, just being someone to listen to someone else. But, for *me* . . . ? No. [laughs]."

Similarly, Susan stretched her affect a bit in order to include the possibility of "helping someone else," but ultimately did not perceive her experience in terms of gifts or generativity:

> A lot of uh, yeah, just a lot of trouble. I think um, you know in thinking about talking to you, . . . I was thinking that *perhaps* it might, um, you know, be able to contribute toward helping other people. I also thought that perhaps it might touch some doctor or some practitioner to . . . to have some compassion or some empathy.

For her part, and in response to a host of symptoms that did not fall neatly into one of the conditions managed by the clinic, Susan had developed a fairly elaborate skin and self-care regimen over the course of several months, and she and her husband had been struggling to maintain physical intimacy. She told me that he was very supportive, that sex was a very important part of their relationship, and that they were both looking forward to things being normal again, which, for them, included enjoying their recently achieved empty nest. Thinking about the care and attention that she was devoting to her vulvar body each day, I asked Susan if she imagined that she would maintain any of her regimen should her symptoms resolve. Her reply was succinct and unequivocal: "*No*. I don't want to have to do anything special. I guess I think that if you're a person that takes care of yourself, you shouldn't have to."

And finally, there was Clair, who on her first visit to Dr. Erlich complained of ten years of vulvar pain ("it feels like someone squirted lighter fluid up there and lit a match") *in addition to* a lifetime of anorgasmia. Clair was reluctant to ascribe any kind of silver lining to her pain and answered my questions about generativity in the "deprived" terms quoted earlier. Clair did not elaborate this sentiment on her own; it developed when I asked whether she was looking forward to the sex that she and Dan were possibly going to have after her upcoming vestibulectomy.[4] Clair answered me with a burst of optimism, which was quickly punctuated with the deprivation and sexual despair that colored the broader experience of her symptoms:

> CHRISTINE: What about thinking about having sex, what does [that] feel like?
> CLAIR: Oh, it's exciting. It's like being a newlywed again, it's, like falling all over, like *falling in love* again. All over again.
> CHRISTINE: So you still have a lot of hope? A lot of optimism?
> CLAIR: Yeah. I have a lot of expectations and I nee—, I don't know what's reality and what's, you know, gonna be normal. . . . And I would, if I go through the surgery and I come home and months goes, and I'm all healed and it's still . . . (trails off). I, I'm going to be very disappointed, and very delusioned [*sic*], and very . . . I'm gonna feel . . . ripped off. Probably.

When I returned to the clinic the week after driving out to Clair's house in the country, I told Dr. Erlich about our interview. Clair had presented us with a dilemma when we'd met her several weeks earlier: in relating her history of anorgasmia, she asked Dr. Erlich if there wasn't something that she could give her "for the desire," something that would make her pain feel less prohibitive. That is, Clair believed (and expressed to us) that if she felt "enough [sexual] *desire*," it would allow her to transcend her pain and enjoyably engage in the kind of ravenous sexual passion that she was convinced "normal" (orgasmic) women routinely experienced. Dr. Erlich and I gently explained to Clair that even if such a treatment existed, we would not offer it to her under those conditions, as no one would (clinically) advocate that she have sex in pain. Clair was disappointed, insisting that the pain would be transformed in the context of *that* kind of desire—the kind that she elaborated to me over our interview, the kind that distinguished "making love" from "having sex," and the kind that would have banished her pain to the margins of a sexual experience that came straight out of a romance novel.

I told Dr. Erlich that Clair and I had again discussed this issue and that I had stressed to Clair that this was how insurance companies rationalized their unwillingness to pay for VVS-related vestibulectomies (i.e., that in the absence of visible pathology, a "functional" vagina was a patent vagina, regardless of subjective accounts of pain), a procedure that Clair herself was scrambling to get covered. Dr. Erlich listened to my frustrated account, and, as we pondered the problems that Clair wanted her surgery to "fix," I thought about not only the ambiguity with which patients and physicians often "live" vulvar pain but also the related dilemma of how feminist scholars might respond to the diagnosis and treatment of "female sexual dysfunction" (FSD). Though a thorough discussion of FSD is beyond the scope of this book,[5] it bears noting that the category has been subject to a fair amount of scrutiny. Whether or not a "pink Viagra" is or should be on the medical horizon is a question driving a good deal of feminist scholarship (Tiefer 2003a, 2003b, and 2006; Potts 2008), while clinical researchers and pharmaceutical companies are beginning to frame the diagnosis in terms of female empowerment and sexual equity (Berman et al. 2003; Berman, Berman, and Bumiller 2005; Canner 2009). Feminist FSD research often conflates vulvar pain patients with women whose anorgasmia or "low" libido are, they argue, profoundly entangled with phallocentric heteronormativity; the "medicalization" of these symptoms is viewed as cynical "disease-mongering" in a culture of retail medicine (Tiefer 2010).

While I share the critical concerns of much of this research, my work with vulvar pain patients has led me to believe that clinical medicine has a legitimate role to play in the management of their symptoms (though by now readers should be aware that I do not believe it to be an all-encompassing or nonreflexive one). The problem, as I see it, is that if genital pain is understood by critical sexuality scholars *solely* as one point among many along a spectrum of disputed disorders, we will miss the opportunity to learn how experiences (like pain) are simultaneously material and discursive, how pain or altered orgasmic capacity can be physiological, affective, and political. Wilson contends that "feminism can be deeply and happily complicit with biological explanation" (2004, 13–14); not only do I agree with this claim, but I believe that we can neither understand nor help women like Clair without such complicity. A feminist dismissal of the biologically "real" clinical needs of women in pain risks duplicating the ways that many of these women have been dismissed by a patriarchal medicine that has not adequately listened to their concerns; it also risks aligning with husbands like Anharrad's and Sharon's, who would prefer to have not heard from their wives' complaining vulvas if it meant compromising their access to penetrative intercourse. Vulvar pain may indeed constitute a symbolic refusal of the heteronormative order of things and, in this way, medicine has little to offer it. But if we do not fully acknowledge the biological dimensions of genital/sexual pain, we are missing the opportunity to understand the mechanisms through which that refusal is processed in and by the body.

7

EVALUATION

Concluding Thoughts

It must be said from the outset that a disease is never a mere loss or excess—that there is always a reaction, on the part of the affected organism or individual, to restore, to replace, to compensate for and to preserve its identity, however strange the means may be.
—Oliver Sacks, *The Man Who Mistook His Wife for a Hat, and Other Clinical Tales*

Then I can stop being an ugly cunt, a silly cunt, a dumb cunt, because people will no longer have to pretend that the vagina is a trivial, idiot, repellent, and inarticulate organ.
—Joanna Frueh, "Vaginal Aesthetics"

Experiences of the vulva are often paradoxical: genital labia are routinely eclipsed—including by gynecological medicine and so-called sexuality experts—by the vagina and its relationship to coital sex and reproduction. In pain, however, these lips that have been pressed into silence begin to speak—sometimes scream—volumes about what female sexuality might (also) involve. Female genitalia that cannot be penetrated, yet remain amenable to caress, massage, and therapeutic attention, elude and exceed received assumptions about (hetero)sexual bodies. These disruptions significantly inform, indeed perfuse, the lived experience of vulvar pain. Patients and physicians must use the space of the clinic, and the alternative verbal and bodily exchanges for which it makes room, to confront and negotiate the *dis-eased* landscape upon which afflicted women more routinely tread.

The previously stifled and inchoate nature of my informants' narratives is what shaped the analytical structure of this book, a book that emphasizes women's experiences of their genital bodies over the structure, political economy, or social location of the "clinic," a typical site of ethnographic engagement (Karkazis 2008; Latour and Woolgar 1986; Livingston 2012; Mol 2002; Saunders 2008). Had I written from that perspective, this book would have included a deeper analysis of many of the actors described in these pages and, more specifically, how they worked together to make the diagnosis of vulvar pain cohere. For example, I might have described the physicians' referral practices and how, despite not knowing much about what physical therapists actually do with their clients, Drs. Robichaud and Erlich dutifully ensured that symptomatic women pursued this line of treatment. The physical therapists (on the receiving end of these underinformed referrals) accumulated more patients, whose bodies subsequently allowed them to improve their clinical skills. The patients benefiting from these sessions would report their progress to the physicians, who would then use that anecdotal information to rationalize continued referrals. Repeated referrals and successful outcomes would help to get insurance codes on the hospital and PT billing forms, which would lead to a greater likelihood that a woman's insurance company would pay for her sessions. For her part, Dr. Erlich would keep checking the box for *pelvic floor myalgia* (rather than *vaginismus*) when ordering PT, which would then help to redefine the discourses through which vulvar pain became medically apprehended. And clinicians with the resources to do so would conduct research about the efficacy of PT, aware of the need to do so in an "evidence-based" medical climate (Feinstein and Horwitz 1997). In this type of analysis, each of these pieces would be analyzed regarding its place in the contemporary state of vulvar pain conditions, each role delineated so as to better apprehend the orchestration of VVS and *vulvodynia*.

An ethnographic account of the clinic might also include my notes from surgery. I could describe how after each procedure I observed, Dr. Erlich would deposit tiny pieces of excised vestibular flesh into formalin-filled cups and then instruct the circulating nurse, who was waiting with a Sharpie, on how to label the specimen. "Oh, is it being sent [for biopsy]?" "Yes," Dr. Erlich would reply. "What is it?" the nurse would ask. "A posterior vestibulectomy." "A what?" "Vestibulectomy," Dr. Erlich, meeting my gaze over our surgical masks and rolling her eyes, would restate the word, spelling it out when necessary. My field notes from one of these mornings say:

> When [Dr. Erlich] cuts off the specimen . . . , she lays [it] carefully on a Telfa pad. She inspects it, sort of lovingly. I gaze at it. It seems so insignificant, incapable of causing this much disruption in a person's life. I think of [Dr. Erlich] in a visit, saying, "This tiny area is what's changing your life." [In this moment], it is not impossible for me to believe that it IS responsible. I want to believe that [she] is right.

In an instant like this, actors, medical supplies, language, and bodily flesh cohere to produce the condition of VVS (Knorr-Cetina 1999; Latour 2007).

The work of this moment is both disrupted and substantiated in different moments, such as when I am changing out of my scrubs in the women's locker room and find myself engaged in conversation with Jennifer, one of the (previously) Sharpie-wielding nurses:

> I say hello and then ask her what it's like to watch those surgeries. "You've seen a few of them now, right?" She says, "Yeah," and seems unimpressed. I start to be disappointed that this isn't something she is all that interested in. And then—BAM!! She tells me that one thing she notices is when they are coming out of the anesthesia, "and that's when I think the real personality comes out," that they are "really whiny". . . . I say, "Oh, 'It

hurts,'—you mean that?" (Libby had complained of a lot of pain during her surgery, from beginning to end.) And Jennifer says, "No, like 'I want a blanket! I'm cold!'" (She says this in a mocking, whining tone.) She says that most patients, when you say, "'Okay, [the blanket] is coming,' they get it. Okay. But these patients, they keep asking, even when you tell them it's coming." And then she says, "That's when you start to wonder what's going on with these patients, whether it's all in their heads, really." (Field notes)

Jennifer tells me that she had been in a relationship where "some of this was going on," and has since wondered "if it wasn't just my body trying to tell me that the relationship was bad. That the man was wrong for me." This nurse's claim that her own pain was resolved when she extricated herself from a "bad" relationship contradicts Dr. Erlich's narrative about VVS, but not necessarily Dr. Robichaud's. Nor does it contradict the story told by some of the physical therapists treating their patients. My purpose in this book has not been to discover or substantiate the truth of any of these accounts but rather to demonstrate one particularly fruitful facet of an ethnographic approach, that is, how the "reality" of a social fact (like a disease) is contradicted and destabilized by the same institutions, actors, and discourses through which it is simultaneously realized.

In this other, hypothetical version of the book, I might have also elaborated some of the less clinical conditions of possibility through which vulvar pain diagnoses emerge. These include the ambivalent nature of postfeminism and the widespread dissemination of sexually explicit media that I have alluded to only briefly. Both sustain a socially sanctioned space for the sexually assertive discourse that allows physicians—if not patients—to directly address the sexual consequences of vulvar pain. These frank discussions articulate with the cultural "moment" of sexual medicine as easily as they do with the popular appeal of television shows like *Sex and the City* or *Girls*. And for a variety of reasons—including cosmetic labiaplasty and anti-FGM (female genital mutilation) campaigns—the malleable and vulnerable vulva is culturally available in novel and overlapping ways. And in a queer twist, the reality of vulvar pain is also informed by the United States' same-sex marriage debate, a connection made clear to me one morning when Dr. Robichaud told me about her patient JoJo and the numerous (and embarrassing) conversations with insurance company executives she had to endure as she tried to secure coverage for her vestibulectomy. We had just been discussing the anti-gay-marriage referendum that was to be on the state's electoral ballot that year and how its "One Man, One Woman" campaign slogan invoked the vaginal-penile intercourse that JoJo was trying to access via her surgery. What, we wondered, was the insurance company's relationship to this tenet of so-called natural law? Did their reluctance to pay for JoJo's vestibulectomy constitute a queering of her marriage, in that it kept her and her husband from behaving heteronormatively? Wasn't penetrative intercourse the mark of so-called normal marriages?

Finally, I might have used ethnographic inquiry to find out, as many of my interlocutors have asked, whether vulvar pain has been there "all along," or, in contrast, I might have argued away this very question. In the first analysis, the disparaging discourses that constrict a symptomatic woman's ability to adequately recognize her pain, that perhaps even contribute to a worsening of her condition, would be understood as inscriptive discourses written onto a "raw" body in real physiological pain. In the second, I might have argued that unmanageable vulvas are "natural symbols" (Douglas 1973) of the discursively "painful" state of female sexuality, and that these recalcitrant symptoms perform the vital (and embodied) cultural work of alerting the social body to an unacceptable state of affairs. In this (second) case, the bodies in question would become hieroglyphs or spectacles through which cultural values and social orders can be interpreted, or discursive and transcendent texts that represent the meanings ascribed to *particular* bodies in *particular* historical and cultural domains.

My arguments in this book are influenced by all of these perspectives, but my work with vulvar pain patients and their providers took my analysis in a slightly different direction. If we posit cultural discourse as a primarily complicating—or exacerbating—force, as it is in the first approach, we run the risk of underplaying the material effects that ideology and social processes generate. But we take an equally dangerous analytical gamble when we overly textualize the physical experience of symptoms, and using a framework that keeps these two approaches in dynamic dialogue does little to demonstrate *how* it is that culture comes to be embodied, either as intensifying cofactor or as interpretable text. In locating my analysis within the body itself, I attempted to resist this binary with questions about culturally coded layers of flesh, discourses that bypass the brain and central nervous system, and the physical recuperation of body parts stolen by cultural narratives.

The expansive nature of fieldwork also allowed me to better understand "how an 'experience' came to be constituted in [a] modern Western societ[y]" (Foucault 1990b, 4), and I invoke Foucault here to formally—and finally—situate this book. In this second volume of *The History of Sexuality*, Foucault continues: "What I planned . . . was a history of the experience of sexuality, where experience is understood as the correlation between fields of knowledge, types of normativity, and forms of subjectivity in a particular culture" (4). My version of Foucault's plan resulted in the chapters that are now behind you, a document perhaps best characterized as a cultural analysis of the vulva—via contemporary and emergent pain conditions. My insights about this overly signified body part, as well as about female sexuality, gynecological medicine, and cultural physiology, are eclectically derived but could not have been gleaned without a methodology that directly involved my own body. It was vital that I sometimes had to drive hours down the coast or over the mountains for interviews in order to physically appreciate the distance traveled by these despairing women. I needed to be invited into Nikki and Sage's home—and see their color-coded closet—so that I could more fully apprehend the material rigidity and order through which they lived their lives. I needed to talk to Jennifer in the locker room about her "bad" relationship, and I needed to pull up to Clair's house and see the sign planted in her front yard, a sign that read "Dwayne's Hideaway" (her husband's name), in order to think more carefully about the *presence* that her pain allowed her to command in that marriage.

Vulvar Disease is everywhere, and it articulates with the multiple and contradictory discourses indexed by these snapshots. Foucault introduces his project for *The Birth of the Clinic* (1973) by stating that "what counts in the things said by men is not so much what they may have thought or the extent to which these things represent their thoughts, as that which systematizes them from the outset" (xix). I have been paying attention to what gets said about female sexuality for a long time, and, via their vulvas, my informants have suggested that this sexuality is vulnerable, ambivalent, erratically subject to heteroregulation, precarious, alienated, and chronically at risk of disappearing. I have used these pages to suggest that vulvar pain—and the sexuality it indexes—cannot, therefore, be successfully "managed" without a full accounting of its collective and cultural dimensions. But because the means through which clinicians can accomplish this task remain underexamined, I present some of the interdisciplinary engagements precipitated by my research, beginning with a final ethnographic anecdote that can frame the discussion.

I have, and I would continue to characterize my relationships with the physicians at Riverview as both productive and mutually respectful. But a comment that Dr. Robichaud made toward the end of my fieldwork reminded me that the nature of my anthropological interventions and analysis remained elusive to them. We were discussing what seemed like a dependence that some of her patients seemed to be developing on me, evidenced by the disappointment that one of them had expressed when I was not able to attend her appointment earlier that week, and we briefly pondered how that reflected our respective goals for these patients. Noting our contentment that my presence at the clinic was at least *enhancing* (rather than negatively complicating) their experiences, Dr. Robichaud remarked, "Well, if you think about it, Christine, how are you different than a therapist?"

In that moment, my feelings were quite mixed: I felt compelled to wonder about my skills as an anthropologist, or at least about my ability to translate my research questions and interests to the professionals for whom my findings would be most relevant. Without any of the requisite education or degrees to qualify me as a therapist, I could assume only that Dr. Robichaud was referring to the support and attention I was lavishing on some of her patients. But had four years of doctoral-level education made me no more than a good—and therapeutic—listener? While I do not disparage this quality, as I believe that it did play a role in the warmness with which I was welcomed into many of my informants' lives, I want to address what *else* I provided to Dr. Robichaud and Dr. Erlich's patients, particularly since I believe that these interventions work as a possible bridge between the disciplines of critical medical anthropology and clinical medicine. That is, in clarifying what I offer(ed) to vulvar pain patients, I can more effectively help readers translate my interventions into treatment options, research agendas, and policy proposals.

It is true that I listened attentively to the stories of diagnosed women. But what is also true is that I was *affectively invested* in their well-being and recovery, a quality of anthropological engagement that can be brought to the forefront of our work, particularly in interdisciplinary contexts. Far from disengaged or alienated social scientists, anthropologists typically care deeply about the people with whom we work, though this does not make us therapists. Indeed, the "processing" that I did with clinic patients was rarely individualized or psychological in nature; rather, our conversations were always guided by my interests in the anthropological and feminist implications of their experiences, and my informants were both willing and able to speak about their lives in these terms. Indeed, I contend that it was this affective investment that enabled me to zero in on the "safety and support" issues raised by Drs. Harlow and Stewart (see chapter 5) and to insist that they be reconciled with the previous shelving of diagnosed women's sexually abusive pasts. Further, these issues take into account the myriad cultural forms and locations of *unwanted genital experience*, which I argue is a substantial component of these disease conditions. Regardless of the physiological, molecular, or immunological markers increasingly affiliated with vulvar pain conditions, I insist that the sociocultural processes described in this book are at least correlated concerns and that, as such, they should be considered alongside other variables as potential "mechanisms" by which adult-onset vulvodynia develops (Harlow and Stewart 2005, 879). Moreover, these processes can be apprehended in constitutive as well as complementary terms; in such an analysis, interventions must be theorized and carried out at the social and political levels through which vulvar pain is experienced. School health and sex education programs that address genital health and well-being, feminist analyses of labiaplasty and of the increasing distance between contraceptive methods and female genitalia, and political alliances between symptomatic and other genitally dis-eased women are just a few examples of these more collective modes of intervention.

Second, the work being funded by the NIH (catalyzed by the NVA's agenda) must include social science–based and humanities-based research. Again, my ethnographic data suggest that even if the conditions of vulvodynia and VVS are purely physiological, the similarities between the demographics and social locations of diagnosed women must be analyzed regarding the corollary, causative, and constitutive nature of these physiological factors. Federal and interdisciplinary support can bring the concerns raised in this book to the same table from which Drs. Foster, Harlow, and Stewart are formulating physiological hypotheses. Such conversations can help us converge our goals, as it should be clear that increased verbal facility and bodily knowledge can help women to more accurately describe the nature of their symptoms. At the least, this can cut down on the number of years between symptom onset and accurate diagnosis, a span of time that now averages five to seven years. Large-scale surveys that include a diversity of racialized, socioeconomic, and sexual situations could help to formulate new research questions about how the social sciences can better understand—and *treat*—the totality of these conditions. Medical anthropology's focus on the political, economic, and other structural contexts of suffering and affliction can help to illuminate how and whether conditions like vulvar pain gather what Michael and Rosengarten (2012) refer to as "medical traction" (6).

Third, providers must communicate with one another about their clinical impressions and hypotheses, as well as about the content of and rationale for their treatment plans and sessions. My data suggest that some physical therapists are not keen on surgery as an effective therapeutic strategy, as many of them believe that successful pelvic floor work can treat most of the problem. Physicians, for their part, don't always understand or intellectually support the bases for some of the approaches taken by physical therapists. I spent a good deal of time thinking about this after I heard Drs. Erlich and Robichaud questioning the value of craniosacral therapy in the hall one morning, as I knew full well that Cathy and Hanna—two of their favorite physical therapists—sometimes employed this technique. These epistemological gaps must be addressed, although further ethnographic research can first provide more convincing evidence of their existence. As this was not a set of questions that I pursued in my research, I have only anecdotal data to support my assertion.

Physical therapists must also be included in the next "state of the art" conference about vulvar pain, as should social scientists. Again, whether or not these disease conditions are purely physiological, we must acknowledge and explicate the cultural milieu in which they occur, as it has profound implications for women's relative abilities to comply with treatment regimens and recommendations. As a clinic that has already welcomed an anthropologist into their midst, the VHC can stand at the forefront of these vital and interdisciplinary collaborations. Einstein and Shildrick (2009) argue that one problem with widening the interdisciplinary net in this way is that health care providers "engage in their craft under certain practical constraints," some of which can "inhibit radical change" (296). Furthermore,

> The urge to impose order," they continue, "to neatly distinguish between good and bad interventions[,] . . . can lead to these practices remaining unmarked by intellectual developments elsewhere that problematize such normativities. In particular, given the practicalities of practicing medicine, it can be argued that operating

from an attitude of contingency and interconnectedness might paralyze the practitioner with possibilities. (296)

Despite these potentially contradictory aims, these researchers laud the merits of "bringing theory back into women's health practices" (Einstein and Shildrick 2009, 298), a goal that might be furthered via creative coalitions. Such an approach may help us to better address some of vulvar pain's more elusive and "slippery" dimensions, such as its religious and racialized aspects, both of which remain relatively undiscussed in the literature. And though both are ripe for critical analysis, the issue of race is comparatively salient given the almost inverse racialized proportions of women with pelvic and vulvar pains (see chapter 2). Ethnographically exploring both the nature and the substance of clinical consultations for each of these conditions, paying particular attention to the language used by both sets of actors, as well as the scheduling and content of follow-up care, can provide us with important new insights about how "black" and "white" female sexual bodies are clinically apprehended. In the preceding chapters, I have argued that women with vulvar pain face tremendous obstacles in their relief-seeking efforts, intangible impediments shaped by a broad and pernicious cultural dis-ease with the female genital body. But because these patients are privileged, rather than marginalized, by the multifaceted social structures of race and class, the struggles they face in accessing care are missed by analyses that stop at the more measurable variables of skin color and income. In a comparative analysis, where "white" and "black" pains are juxtaposed, this argument can be extended by attending to the possible hurdles faced by women of color as they attempt to be diagnosed with a "white" pain.

The hypersensitivity discourse within vulvar pain circles merits further interrogation as an apparent by-product of this community's efforts to medically legitimate their formerly psychosomatic pain. Many of the physicians who looked harder and further for the physiological bases of female genital pain have settled—thus far—on neurological explanations. Theories of "fired up" nerves now pepper the clinical literature, driving the development of treatment options meant to calm them

(De Andres et al. 2015; Foster, Dworkin, and Wood 2005). As I have described, this framework comes dangerously close to nineteenth- and early twentieth-century racist and eugenically inflected discourses that equated such delicacy with an elevated level of civilization, relegating "coarser" women—working class, nonwhite, immigrant—to lower rungs on the evolutionary ladder (Horn 2003).

For slightly different physiological reasons, women with both vulvar and pelvic pain cannot participate in routine penile-vaginal intercourse and, in this way, also trouble the patriarchal order of things. Since Alfred Kinsey began asking them, US women have consistently reported that conventional penetrative intercourse is not their preferred route to orgasmic satisfaction, regardless of how much pleasure they take from it. Despite several waves of feminist activism and at least one so-called sexual revolution, however, heterosexual women struggle to find partners who will consistently explore alternatives (Potts 2002). This book argues that this aspect of vulvar pain is a potential site for feminist (hetero)sexual theory, because bodies that both desire and refuse masculinized sexual scripts may index important ambivalences and confusion in a "post-feminist" era. A comparative analysis with chronic pelvic pain could both expand and deepen this argument by adding a second group of voices to the data I have gathered thus far.

In its broadest sense, such a comparative project can question the nature of medical categories and disease classifications and interrogate the embodied and epistemological filters through which particular bodies are understood. Clinical medicine operates well within the discursive and material lines dividing black and white female bodies in the United States; the specific conditions under investigation in this project index, rather than constitute, disparities that cannot necessarily be remedied with greater material "access" to care. In the VHC, many African American women walk through the doors of the Women's Health Center, but turn left (toward the resident clinic) at the place where (white) vulvar pain patients walk straight—through a different set of doors and into a waiting room far less crowded and uncomfortable. As I have argued, I suspect that the physical and emotional burdens experienced by both groups of women in confronting their sexual pain are more similar than disparate and that their narratives can mutually inform the bodies of literature from which feminist and other disparity theorists formulate their interventions. But the physical segregation of these bodies confounds our abilities to make these connections. Research building on this book can configure new lines of connection and ethnographically substantiate the provocative lines of similarity drawn by Harvard's researchers in their efforts to disrupt the settling demographic profile of women with vulvar pain. In so doing, I hope to locate new sites from which feminists, medical anthropologists, and clinicians can better understand the nature of the stories that "raced"—as well as gendered, classed, and other categorized—bodies continue to tell.

SO WHAT *IS* VULVAR PAIN? IS IT REAL?

Irigaray (1985b) suggests that a discourse's coherence is ensured by its uninterpretability. In this book, I have sought to interpret—and therefore destabilize—how heteronormativity coheres around vulvar invisibility, in order to offer a new set of tools with which to (re)think the female genital and sexual body and to broaden the limits of what counts as a valuable body in the world. But in following this course, I may have left readers with nagging questions about what vulvar pain *is*, exactly, or more precisely, why only some women are afflicted, and what all of this might mean regarding genital health and illness.

Because I approach these conditions as an anthropologist, I understand them in biocultural terms, that is, as somatic experiences thoroughly entangled with the discourses and practices through which women come to understand their sexual anatomies. That said, my reading of the clinical literature regarding vulvodynia and VVS (also known collectively as *vestibulodynia*) suggests that they are also physiological pain conditions to which some women are more predisposed than others. My research—which brings ethnographic engagement with patients, partners, providers, and physical therapists to bear on this literature—suggests that the risk for adult-onset vulvar pain is increased for women whose childhoods have been particularly onerous and who have collected an inordinate amount of unwanted genital experience. At times this experience has been in the form of explicit abuse—sexual, physical, emotional, or otherwise—while at others it was in the form of proscriptive (religious) doctrines through which sex, genitals, and marital behavior were severely compartmentalized. For someone like Nikki (see chapter 4), whose history included all of these variables, VVS was an almost predictable development—Harlow and Stewart's 2005 data estimated that the risk of adult-onset vulvar pain for women growing up in households like hers was fourteen times that of women with safer and more supportive pasts. And though the precise mechanism through which her pain developed is still far from clear (NIH 2012), the course that it took, including a series of structural and individually inflected delays, as well as its nature and severity, can be comfortably explained by physical therapists with expertise in vulvar pain.

Given these data, along with the availability and increasing reliability of screening instruments (Harlow et al. 2009; Reed et al. 2006; Reed et al. 2012), it seems both possible and beneficial for clinicians to include vulvar pain assessment in their delivery of primary care. Indeed, the fact that such screening mechanisms are the principal way through which women of color have found their way in to prevalence statistics makes the widespread adoption of this practice necessary, not only for more expedient care delivery but also for enriched clinical descriptions of the conditions. To that end, heightened awareness of vulvar health should be directed toward *all* people with vulvas, including trans men and other queer and gender-non-conforming people, some of whom may not present as women to their health care providers (Grant et al. 2011). Whether and how these populations experience vulvar pain can inform precisely the kinds of questions with which this book is preoccupied, questions that emphasize how genital pain conditions are obligated to their respective cultural and historical milieus (Karkazis 2008; Valentine and Wilchins 1997).

Following Wilson (2004), I have employed the concept of obligation in order to characterize vulvar pain. Reducible to neither physiological nor discursive conditions, vulvodynia and VVS are simultaneously biocultural and infrastructural, multifaceted and dynamic entities that precipitate and amplify one another's effects; this is acutely evident in the ways that vulvar pain is still "missing" from public conversations about health. Without a reliably effective treatment or pharmaceutical agent to monetize and stabilize its reality (Dumit 2012), vulvar pain remains quasi-legitimate in clinical terms. And though this can intensify the personal distress of symptomatic women (whose partners and friends don't believe them, for example), the lack of a "go-to" treatment option *also* makes their extended—and even elaborate—suffering more permissible than that of patients for whom effective drug treatments are allegedly available.[1]

But regardless of effective risk profiles, screening mechanisms, or treatment regimens, vulvar pain will continue to be exacerbated if and when a woman is unable to describe her symptoms, thinks that her symptoms or her genitals don't matter, or has been led to believe that vaginal intercourse necessarily includes some amount of pain. It will also be worsened by inattentive providers, and by institutional practices that carve out diagnostic entities based on skin color, age, or sexual habits rather than by lived and bodily experience. For some cultural theorists, vulvar pain can be understood as an assemblage—a "world" (Stewart 2010) crafted by and through the various actors and elements outlined in this book: symptomatic women, partners, health care providers, popular media, insurance companies, health columns, vaginal dilators, websites, romance novels, pharmacists, pornography, well-meaning friends, surgical tools, sex therapists, academics, joke-tellers, and conference organizers. These are the components of a (female) structure of feeling that is disinvested in its genitalia; interpreted in their collective form, they constitute compelling evidence for a more thorough accounting of the health consequences of culturally mediated bodily shame, particularly as it contributes to an inability to communicate with partners or providers. It is this world to which I refer with the capitalized phrase Vulvar Disease, a condition for which this book seeks a biocultural cure.

EPILOGUE

Collaboration

———◆———

> Moreover, some women may be reluctant to discuss their pain or seek treatment.
> —National Institutes of Health, "Research Plan on Vulvodynia"

In July 2011, the National Institute of Child Health and Human Development (NICHD)[1] held a two-day scientific meeting entitled *Vulvodynia: A Chronic Pain Condition—Setting a Research Agenda*, the goal of which was to "gather input . . . to guide the field of vulvodynia research" (NIH 2012, 6).[2] Participants included many of the researchers present at the 2004 conference (see chapter 4), who were now joined by new researchers in the field, as well as representatives from the NIH's Office of Research on Women's Health (ORWH), the pharmaceutical industry, academic medicine, and patient and provider organizations. Also present at the meeting were researchers from the National Institute of Dental and Craniofacial Research (NIDCR), who signaled the NICHD's interest in contextualizing *vulvodynia* within a broader spectrum of chronic pain conditions (NIH 2012, 2).

The resulting thirty-page report, the NIH's "Research Plan on Vulvodynia" (2012), describes a meeting that focused heavily on the physiological dimensions of vulvar pain, including

> the history of pain and pain research; the recommendations of the then-recently published [Institutes of

Medicine] report on pain research . . . ; appropriate ways to measure pain for research purposes; potential treatments for vulvodynia based on blocking the action of specific receptors; [and] applying research results on the pathophysiology of fibromyalgia to vulvodynia. (6–7)

The report also details a number of research-related goals that emerged from the interdisciplinary nature of the conference. These included greater precision and universality regarding terminology, increased attention to comorbid conditions (particularly other forms of chronic pain) and the role of hormones, and a greater attunement to the "psychosocial aspects of vulvodynia, especially those related to sexual function" (NIH 2012, 6–7). In its broadest terms, the research plan endeavors to "enhance capacity for conducting research related to vulvodynia, and to address *key areas* needed to move the science forward" (5; my emphasis).

I began this book by parsing the phrasing of another set of clinical researchers, that is, the "other causes" of vulvar pain to which epidemiologists Harlow and Stewart referred in their now-seminal 2003 prevalence article. It seems appropriate, therefore, to conclude my analysis by briefly reflecting on the "key areas" identified by the NICHD, including their decision to locate vulvodynia within the wider field of chronic pain conditions. The research plan asserts that the specialists who gathered in July 2011 "emphasized" (NIH 2012, 1–2) the importance of this move, leading to the short-term goal of "increasing scientific outreach efforts to the broader pain research community to encourage them to apply their scientific knowledge to vulvodynia research" (1–2). The researchers present at the meeting also stressed the short-term importance of more evidence-based diagnostic and treatment plans, greater interdisciplinary collaboration, and developing training and curricular materials for emerging gynecological and general practice clinicians.

In addition to mass screening programs, longer-term goals of the NICHD plan include developing vulvodynia-specific animal models, tissue banks, biomarkers, and phenotypes,

EPILOGUE   253

in addition to mapping the natural history of vulvar pain in research subjects who have not yet acquired the disease. And though the report gestures toward the "psychosocial" aspects of vulvar pain conditions, none of its explicit goals or directives address how these dimensions might be investigated. Indeed, the previous sentence regarding "psychosocial aspects" is one of only four (in the entire document) to acknowledge that the experience of vulvar pain is *also* an extraphysiological one. Part of the problem, the report acknowledges, is that these other effects are less empirically precise and (perhaps) less meaningful: "Scientists stated that sexual health measures were important, but that they were not sufficient by themselves to identify quality-of-life impact. Researchers also discussed the need to examine the relationships among psychosocial factors and the incidence and severity of vulvodynia symptoms" (8).

Though I had initially worried about missing the 2011 meeting, my subsequent readings of the research plan felt less like the clinical update I would have hoped to receive than an experience of déjà vu. From the (re)recognition that physical therapists need to play a greater role in the research process, to the sandwiching of "chronic stress, early life trauma, and abuse" (NIH 2012, 25) between bullet points about nerve blocks and neuronal stem cells (and limiting this discussion to that one sentence), it is clear that the NICHD's understanding of vulvodynia does not include the disease's profound entanglement with the ways that the female genital body is always and already culturally contaminated. And though the research plan claims to want more information about the *kinds of* women (see chapter 3) more likely to develop vulvar pain, it does little to recognize the concomitant "condition" of vulvar *dis-ease* and disparagement, an affliction that will compound the experience of women with a physiological pain condition and serve as kindling for those who might develop one in the future.

As I read and reread this report, I remain frustrated that such a large gathering of obviously knowledgeable people have (once again) missed this vital connection. I am vexed by the apparent reality of an eminent group of researcher-clinicians who either do not or cannot recognize that vulvar pain conditions will be resolved more easily—and more completely—when symptomatic women are full participants in their recovery: when they can name and understand the site of their pain, when they can successfully bring this pain to a clinician's attention, when their partners understand and collaborate in their treatment, and when "fixing the problem" is not defined by never thinking about or referring to their external genitalia again. Whether or not nerve blocks, Botox,[3] or high-tech animal models constitute new benchmarks in vulvodynia research and treatment, they do not address the genitally proscriptive social cues that accumulate in the bodies of women like Nikki, Colleen, Judy, Anharrad, Clair, and all of the other women whose stories I've shared in this book. They do nothing, as I have argued, to alter a cultural landscape from which female external genitalia remain fundamentally absent.

In late 2012, I met with one of the expert clinicians who had attended both the 2004 and the 2011 conferences. Our conversation was relatively brief—no more than an hour—but it was long enough for me to learn that this clinician was "backing off" from vulvar pain work, due partly to shifting professional interests but also because of the difficulties faced in recruiting and hiring new clinicians to the practice. "These patients are really hard," the clinician finally told me. "They're a lot of work and not everybody wants to spend the time it takes to manage them successfully. Even when they come here to do it, they move on to other specialties once they realize how time-consuming and ungratifying it is."

Having now spent over a decade living with the topic of female genital pain, I can say that though I found this clinician's frankness disheartening, I did not find it surprising. New clinicians lose interest in vulvar work for many of the same reasons that the NICHD wants to recode vulvodynia as a pain condition (rather than a sexual one). Confronting—and disrupting—the intransigence of heteronormative sexual narratives that compound the lived experience of vulvar pain requires a kind of "sex work" that is both laborious and rare. Physicians infrequently have the tools to collect adequate sexual histories from their patients (Andrews 2000; McGarvey et al. 2003; Ng and McCarthy 2002; Rosen et al. 2006) and virtually never possess the social scientific skills to analyze how those individual histories articulate with the more pervasive and hierarchical social structures through which women live their (genital) bodies. According to the research plan, "new [and] young investigators" are averse to the "complexity of the condition and the diversity of the disciplines required for significant scientific progress" (7). *Dealing with* these painful vulvas is what clinicians who exit the specialty faster than they entered it often report: a drag—on clinicians' time, energy, ingenuity, and the part of their professional personality accustomed to making a measurable difference in their patients' lives. These are the reasons that women live with unexplained pain for five to seven years and why "getting better" often includes more distancing of the vulva than keeping it close.[4]

In their consensus regarding who should manage vulvar pain, the NICHD participants appear to acknowledge these retention difficulties. Highlighting that "too few investigators . . . , particularly in fields other than gynecology, were sufficiently knowledgeable and *interested in* vulvodynia" (NIH 2012, 7; my emphasis), the report concludes that novel strategies may need to be deployed in order to increase the number of invested clinicians and "move the field forward" (7). "Several participants suggested leveraging existing programs for research training and career development in gynecology to help expand the . . . field" (NIH 2012, 7); others spoke of the possibility of moving vulvodynia research away from gynecology and sexuality and toward the burgeoning field of neuroscience (7), a medical subspecialty with more professional "sex appeal" than actual genitalia (Kempner 2014, 52–53).

But recruiting clinicians to care for afflicted vulvas by making genital pain neurobiological rather than sexual may be savvier than it is sustainable. Though it might, in the short term, generate research dollars and the interest and energy of "young investigators" for whom the brain is more of a symbol of "technological sophistication" (Kempner 2014, 162) than the vulva, it is unclear that this move will offer any relief to the women on whose behalf it is allegedly being made. It is, however, a move in line with the analysis of this book, in that it indexes the ease with which clinicians—and, eventually, their patients—can segregate painful female genitalia from the sexual and bodily experience to which I argue they must remain attached. In other words, relocating vulvar pain to the brain might make it easier for everyone to talk about it, but it does little to reposition female sexual autonomy and bodily integrity away from the tenacious and contaminating realties of cosmetic labiaplasty and general vulvar ignorance.

My skepticism regarding the neuroscience of vulvar pain does not extend to the kinds of dynamic biology around which feminist, anthropological, and other science studies perspectives have begun to coalesce (Downey 2009; Fausto-Sterling 2005; Jordan-Young 2011; Karkazis 2008; Wilson 2004). Indeed, projects like those of Gillian Einstein, who insists that discussions of female genital cutting (FGC) should be informed by the neuroanatomical and sensory details of the vulva and clitoris, are quite attentive to physiology. But, and as Einstein astutely argues, since pain "is a complex perception depending on both neurological substrate and cultural expectation/meaning," understanding the neuroscience of cutting-induced sensations "requires contextual study within women's lived lives" (2008, 94). Asking women themselves what they feel, she continues, "in both a figurative and literal sense" (95), is integral to any project that attempts to explain "how cultural practices instantiate neural circuit differences" (95), a goal shared by a growing number of researchers at the intersections of the social and the life sciences (Losin 2013).

And though my own analysis about vulvar pain was catalyzed by—and subsequently cohered around—Wilson's (2004) discussion of the "maverick" enteric nervous system with which discussions of gastrointestinal symptomatology must, she insists, be reckoned, my conclusions do not rest comfortably on an autonomous and objectively observable genital nervous system. Rather, they remain rooted in the cultural locations, processes, and dynamics with which the symptoms of vulvar disease are always-already obliged. In this milieu, where symbolic and material erasures pile up alongside and between traumatic, contaminating, and disparaging vulvar events, female genitalia are discursively—and *distinctly*—primed to suffer their worlds in particular ways. By fully incorporating the aspects of worlds that clinical researchers would prefer to sideline—without sacrificing the meticulous attention to the body that biologically attuned feminists have placed back on the map—we can more readily conceive of something like a culturally charged vulva. Alternately innervated and anesthetized by shame, hypervigilance, and violence (both symbolic and material), this vulva indexes a female sexuality that exceeds the boundaries of heteronormativity and reproductive capacity, offering female-centered exit strategies to those who can *incorporate* its unsettling potential.

NOTES

PROLOGUE

1. http://talkingpointsmemo.com/news/vagina-sculpture-traps-us-student.
2. http://www.sfgate.com/news/article/Thoughts-Freud-Vagina-sculpture-traps-US-student-5572495.php.
3. http://mashable.com/2014/06/23/student-stuck-in-vagina-sculpture/.
4. Given that its walls touch one another when not being penetrated, the vagina is often defined in these terms.

CHAPTER 1. INSINUATION

1. In addition to my doctorate in anthropology, I hold bachelor's and master's degrees in nursing, and, for almost fifteen years, I worked as a nurse practitioner specializing in gynecological medicine.
2. I will use this highly imperfect phrase throughout the text in order to make a series of cautious generalizations. Though my fieldwork was conducted with a small and narrowly defined group of women (primarily white, insured, straight identified, and well educated), the claims I make about "many women" are also based on the fifteen-plus years that I practiced as a gynecological clinician, during which

I cared for a much wider diversity of patients. Moreover, I recognize the numerous women in the United States (and in the world) for whom genital reluctance is not a problem; indeed, many of the women cited in these pages (and from whom I have learned) frankly acknowledge and often celebrate their genital anatomy. In the larger US context, however, I believe that these women constitute the exception rather than the rule. This book is about the experience of vulvar disparagement that I believe at least partially structures the genital realities of a majority of US women.

3. All of the names in this book are pseudonyms, the majority of which were chosen by my informants.
4. Jackson is citing Schutz (1972) and Kessler and McKenna (1978) with this concept.
5. As with the phrase "most women" (see chapter 1, note 2), my use of the term "female" is also imperfect. While my definition of the term includes anyone who identifies as such, my attention to vulvar anatomy and experience excludes some of those people. Though my research and political interests are attuned to the vulvas of trans women, this study did not include any, and it remains unclear whether women whose vulvas have been surgically acquired are at risk for the conditions described in this book. See Einstein (2008) and Karkazis (2008) for critical discussions of female genital neurophysiology.
6. Gatens is paraphrasing Freud here. The full quotation is as follows:

> This combination of circumstances leads to two reactions, which may become fixed and will in that case, whether separately or together or in conjunction with other factors, permanently determine the boy's relations to women: horror of the mutilated creature or triumphant contempt for her. These developments, however, belong to the future, though not to a very remote one. (Freud 1925, 336)

7. Second-wave feminism has been characterized in several ways, often as a movement that, in foregrounding the similarities among women, downplayed and at times ignored the

important differences—particularly of race, class, sexual orientation, and physical ability—that shaped their respective experiences. The women's health movement through which many US women became aware of the power dynamics between male physicians and female patients occurred during this second wave. See Clawson (2008), Findlen (2001), Jacob and Licona (2005), and Schulte (2011) for further discussion and clarification of the "waves" of feminism. See the Boston Women's Health Book Collective (1992) for an exemplary product of the women's health movement.
8. http://www.someecards.com/usercards/viewcard/a09289 09009e03d48a39b8e950183cf4.
9. Also referred to as sexual reassignment surgery (SRS). See Feldman (2008) and Pitts (2009).
10. The models were large enough to be used as kayaks and were (unsurprisingly) reported by news outlets to be in the shape of her vagina. http://www.bbc.com/news/world-asia-28323015.
11. A psychological phenomenon known as "priming the unconscious." See Carey (2007) for further discussion. See also Xu's (1999) discussion of *qigong* in contemporary China and the assertion of some masters that they can release their own *qi* into another room for the benefit of others (978).
12. http://store.vajazzling.com/.
13. http://www.thefrisky.com/2012-01-12/the-top-9-most-amazing-vaginas/.
14. http://www.vulture.com/2008/01/ten_movie_vaginas.html. *Teeth* is a 2007 horror-comedy film, written and directed by Mitchell Lichtenstein, about a teenage girl with *vagina dentata*, a mythical condition in which the vagina is filled with (castrating) teeth. See Jim Emerson's 2008 review on RogerEbert.com, which, though positive, repeatedly refers to the protagonist's vagina in "down there" terms. http://www.rogerebert.com/reviews/teeth-2008.
15. http://www.mynewpinkbutton.com/.
16. See also Gilman (1985), Gould (1985), and Schiebinger (1993).
17. A growing, and excellent, feminist literature exists on this

topic. See especially Braun (2010), New View (2000), and Tiefer (2010).
18. It is important to note that, like much cosmetic surgery, labiaplasty is not a worldwide phenomenon. Though beyond the scope of this book, there are important similarities and differences between these relatively new and, for lack of a more precise term, "Western" procedures and the kinds of genital cutting that have long been practiced by other cultural groups. For an excellent starting point, see Bell (2005), Braun (2010), New View (2000), and James and Robertson (2005).

CHAPTER 2. EXAMINATION

1. Consistent with my informants' pseudonyms, I have also assigned a fictional name to the clinic, the hospital that housed it, and the physicians and staff with whom I worked.
2. A clinical term denoting the prototypical configuration of objective (signs) and subjective (symptoms) measures of a disease condition; what social scientists might label an "ideal type," as it is often more of an unattained standard than a reality.
3. The change from vestibulitis to vestibulodynia reflects a shift in thinking regarding etiology; *itis* denotes inflammation, while *dynia* reflects pain. Currently, pain specific to the vestibule is considered to be more appropriately categorized as chronic pain than as inflammation (Bornstein et al. 1997; Goldstein and Burrows 2008).
4. SSRI is the acronym for selective serotonin reuptake inhibitors, a class of antidepressant that includes the drugs Prozac, Lexapro, Paxil, and Zoloft. A newer version of these drugs—SSNRIs (Cymbalta)—also regulates the reuptake of norepinephrine. During my time at Riverview, these were just being introduced, and clinicians were hopeful that they could be more effective with vulvar pain patients. Clinical studies thus far have been mixed. See De Andres et al. 2015 for an excellent discussion.

5. These lasers are primarily used by dermatologists to remove birthmarks and other vascular skin lesions.
6. A vestibulectomy removes the entire vulvar vestibule at a depth of just a few millimeters and a width that extends to the line of demarcation known as Hart's line. The surgeon pulls down the skin at the opening/outer third of the vagina in order to replace the excised tissue. In Dr. Goetsch's "modified" vestibulectomy, a smaller area of the vestibule is excised (mapped by the Q-tip test and the patient's subjective reporting of pain) and hymenal tissue, rather than the vagina itself, is pulled down to cover the area. See Goetsch (1996) for further details. See also figure 4.
7. Both of these drugs were used at Riverview but usually with more established patients (i.e., patients for whom they had been prescribed before SSRIs became available). Patients described difficulty with side effects, but those who were taking these drugs reported the best pain relief of any of the vulvodynia patients I saw.
8. A notable one being a "low-oxalate diet." One theory about vulvar pain conditions is that they are related to (or at least exacerbated by) the vulva's exposure to chemical compounds known as urinary oxalates, and that a stark reduction in their intake (and therefore output) can mitigate pain sensations. Oxalates are prevalent in fruits and vegetables, however, making the diet difficult to maintain, especially for vegetarian women. See Bachmann et al. (2006) and Baggish, Sze, and Johnson (1997) for further discussion of this treatment (both pro and con).
9. The NVA is an advocacy group focusing its efforts on lobbying, research funding (primarily hard science), and education. Members and representatives epitomize the demographic profile of vulvar pain—educated, white, and financially stable. They publish a quarterly newsletter, assist with establishing regional and local support groups, and fund one to two small research projects per year; their services are not free of charge, and an annual membership is currently forty-five dollars per year (there is also a "financial hardship" application). Many of the patients I met

chose not to join, although it was the only support group that Drs. Robichaud and Erlich specifically recommended to patients. They typically gave each new patient a brochure, telling them, "This organization is patient-centered and physician-supported." See their website, www.nva.org, for further information.

10. In plant science, the term "lichen" refers to any organism composed of both algae and fungus. In clinical usage, however, it refers to a group of skin eruptions characterized by papules (non–pus containing, raised, and circular lesions).
11. Systemic lupus erythematosus, or SLE.
12. Jenny was Mary's spouse.
13. Savage has since apologized for this remark.
14. See chapter 4 for an extensive ethnographic example of this situation.
15. See, especially, Briggs 2000; de Marneffe 1996; Didi-Huberman 2003; Ehrenreich and English 1973, 1978; Gilman 1985; Showalter 1985; Welter 1966; and Wood 1973.
16. Scheurich is citing Edward Shorter (1992) in his use of the "symptom pool" concept.
17. These are the antidepressants amitriptyline and gabapentin, drugs sometimes prescribed for bipolar disorder, seizure control, and neuropathic pain conditions.
18. Again, see Kempner (2014) for a useful comparison with migraine, in which she addresses "the problematic fact that many of the conditions with which migraine is comorbid, like depression and anxiety, are themselves highly gendered, deeply feminized, and come burdened with their own set of cultural baggage" (49). A "migraine brain," she argues, is likened by physicians to a "highly sensitive" diva (72–74).
19. The hospital has been recognized by the federal government's Office on Women's Health for its innovative, comprehensive, multidisciplinary, and integrated delivery systems of women's health care.
20. Though true at the time of my fieldwork, this is no longer the case.
21. Aside from Dr. Jenkins, one resident, and one medical student, the entire clinic staff during my fieldwork was female,

and a patient could always request that she not be examined or cared for by a male provider, if that was her preference.
22. Though beyond the scope of this book, it is worth noting that the distribution and use of contraceptive methods, including the involuntary sterilization of poor women and women of color throughout the twentieth century, is heavily marked by race and class. See Begos et al. (2012), Kapsalis (1997), Ordover (2003), Roberts (1997), and Schoen (2005) for further discussion.

CHAPTER 3. ACCUMULATION

1. Close to half of the patients that I met had been diagnosed with at least one concomitant autoimmune disorder. The prevalence of these disorders is highest among women, and some scholars argue that they should be gendered female (Ahmed et al. 1999; D'Cruz 2007; Whitacre et al. 1999). Some authors have suggested that vulvodynia and VVS have an autoimmune component and, though a causal relationship has not been demonstrated, there is growing evidence of concomitancy between vulvar pain and irritable bowel syndrome, interstitial cystitis, and fibromyalgia (Chadha et al. 1998; De Andres et al. 2015; Glazer and Rodke 2002; Nguyen, Reese, and Harlow 2015; NIH 2012).
2. Kempner (2014) describes this phenomenon among migraine sufferers, particularly in blog posts and online support groups.
3. Contemporary students are increasingly relying on virtual, or nonbiological, bodies to learn many of these skills. See Prentice (2012) for a thorough recounting of the relevant clinical and pedagogical debates surrounding this transition.
4. "You're hurting me," being one example. The role of professional pelvic model/GTA includes orienting the student to the anatomy being examined ("Yes, that's my ovary") and offering feedback when any part of the exam is uncomfortable or unprofessional in her (informed) opinion. See Jamison (2014).

5. This varied to some degree, based on patients' interest and ability to participate. At a minimum, "involving" her meant keeping her aware of what I was doing "down there"; at a maximum, it might mean dispensing with the drape sheet, keeping a mirror in her hands, and involving her partner in the exam.
6. The word "unremarkable" is used clinically to indicate that a patient, body part, or system is apparently normal. In other words, there are no "remarks" to make.
7. The first anecdote in Kapsalis's (1997) book involves a woman whose physician left her in stirrups after he was called to attend to a more urgent issue. Neither he, nor any other staff member, told her that she could take her legs down while she waited for him to return. Afraid to question his authority, she stayed in this position until he returned, even when she realized that the door was open and there were people walking by. Neither she nor the doctor mentioned it upon his return.
8. Bartholin's cysts are one potential complication of a vestibulectomy. See Goetsch (2009) and Tommola, Unkila-Kallio, and Paavonen (2011).
9. While Dr. Robichaud was telling me about the complication rate for these surgeries, she added that she was performing one the next morning. Having just absorbed all that she'd said to Frances and me, I couldn't help but ask, "My God, are you nervous?" She didn't miss a beat before exclaiming, "Hell, yeah! Especially 'cause I've only done like eight of 'em!" Given the differing investments in the vulvar integrity of their patients, it is difficult to imagine any of Dr. Robichaud's patients ending up like Frances, however, despite the fact that she has performed fewer procedures than the so-called specialist who "cared for" Frances.
10. I now think of this as the penetrative triumvirate, routinely—and unquestioningly—alluded to by both providers and patients. It was noteworthy that the woman's own hand or finger was never included.
11. Think about the difference between muscular tension in your back that hurts only when someone begins to touch it (e.g., massage) versus tension that hurts all the time.

12. I also spent time with two therapists named Sandy and Lisa, both of whom worked out of Riverview and neither of whom used biofeedback. Lisa, who was in the second "camp" of therapists, did craniosacral massage; Sandy—in the first camp—did not. Cathy's group also did craniosacral massage, but not during any of the sessions that I observed. See chapter 4 for further discussion of Lisa's approach.

CHAPTER 4. MANIFESTATION

1. Lisa left Riverview when Sandy returned from her maternity leave, having conducted half a dozen sessions with Nikki. Nikki had approximately that many more with Sandy before I left the field.
2. It is also increasingly standard for this question to be asked by mainstream gynecologists. It often serves to alert the provider that the speculum exam may be a "difficult" one.
3. These patients probably totaled no more than a dozen. Given the available prevalence data (Goetsch 1991; Harlow and Stewart 2003), it is worth noting that this constitutes significantly less than 15 to 18 percent of my total practice.
4. These women and physicians are the genesis of the National Vulvodynia Association (NVA).
5. According to Harlow and Stewart:

    > Sexual abuse was defined as threatened, attempted, or actual infliction of sexual harm. We categorized the source of physical and sexual abuse as primary or secondary family members, strangers, schoolmates, neighbors, or other acquaintances. We also inquired about emotional and social support at home while growing up and the extent to which respondents felt in danger at home, at school, or in their neighborhood during childhood. (2005, 874)

6. The physicians at Riverview were relatively nonconversant about CST. I once heard them speaking about it in the hallway in a somewhat derogatory fashion, conflating it with naturopathic and other modes of health care delivery that

they perceived to be ineffectual. I mention this not to chastise the physicians but to reiterate the peculiar relationship between PT and the VHC. That is, Cathy and her colleagues were beloved by the Riverview physicians; they were also CST adherents. Since the physicians were relatively disinvested in the details of PT, however, this contradiction did not create any cognitive dissonance regarding their PT referral practices.

7. Also a pseudonym.
8. For a thorough discussion of the complex relationship between Christianity, virginity, and female sexual agency, see Carpenter (2005).
9. Pollock is partially citing Krieger (1999), whose sentence refers specifically to a "racist" society.
10. Both are autoimmune conditions that frequently co-occur with vulvar disease (De Andres et al. 2015; Nguyen, Reese, and Harlow 2015; NIH 2012).
11. The term for a vulva that has been stitched together in one of the rarest forms of traditional genital cutting. Judy's urinary complication is commonly cited by anticutting activists, as it has been well documented by a number of researchers and health care providers. Other complications of ritual female genital cutting, most of which are associated with this more extreme, and rarer, form, include: profuse bleeding or hemorrhage, infection, abnormal and painful scarring, menstrual complications, the development of fistulas between the vagina and urethra or vagina and rectum, painful vaginal intercourse, neurological damage, and death. See Dorkenoo (1994), Hosken (1994), James and Robertson (2002), Lightfoot-Klein (1989), and Rahman and Toubia (2000) for a basic overview of the issues surrounding genital cutting as well as further details about these complications.
12. In addition to the risks of a "modulated" immune system (e.g., decreased ability to fight infection), the side effects of these drugs vary from nausea, sleeplessness, and skin irritation to an increased risk for diabetes and some cancers (Numakura et al. 2005; Ormerod 2005).

13. This means prescribing a drug or treatment for reasons other than those for which it has been approved by the Food and Drug Administration (FDA).
14. There is an important—and growing—literature regarding gender, sexuality, and aging. See, especially, Calasanti (2007), Marshall (2006), and Wentzell (2013) for excellent examples.
15. These sessions were the source of many jokes between patients and providers centering on what one was supposed to do while the dilator was inside: Watch the news? Vacuum? Return emails?
16. Because of the area's naturopathic-friendly environment, patients who could not tolerate standard base creams could often access a compounding pharmacist who would mix the raw form of the drug (the modulator) into a blander emollient base, such as petroleum jelly. I estimate that approximately half of the lichen planus patients I met fell into this category. Since these "compounding pharmacies" were primarily an urban phenomenon, women who came to the clinic from distant locations would have their prescriptions filled before they began their trips home, some of which required an overnight stay.
17. This comment might cause us to wonder whether Viagra would be as popular if it required a penile route of administration.
18. Though it is not my intention to suggest that this behavior is male or masculine, each of the three scenarios that were described to me involved a male pharmacist.
19. The first one was titled "Genetic Variation in Interleukin-1 Receptor and Melanocortin-1 Receptor Genes on Vulvar Vestibulitis Syndrome"; the second was "Comparative Effects of Intradermal Foot and Forearm Capsaicin Injections in Normal and VVS Afflicted Women."
20. Partial quotation is from François Frier, *Guide pour la conservation de l'homme*, 1789.
21. The importance of conference-contoured medical realities should not be underestimated. More than one patient recounted how one or more treatments that a previous

doctor had tried with her were based on their having encountered or learned about it "at a conference."

CHAPTER 5. INTEGRATION

1. I use quotation marks around this phrase in order to disrupt its biomedical definition. My use of the phrase, which emerged from the experiences of patients, refers to pain reduction that is obliged to an embodied acceptance of its extraphysiological meanings. See Wolf (2014), particularly readers' comments, for a recent discussion of this.
2. For example, Cathy told me that in response to the question "Which position do you think you want to be in the first time you do it?" almost all of her patients answered, "On top."
3. Through clinical practice, research and publishing, conference presentations, and professional networking, Drs. Robichaud, Jenkins, and Erlich have become established vulvar pain "experts." This is important not only for their own career trajectories but also for patients whose insurance companies balk at paying for a visit to Riverview if the hospital or physician is "out of network"; a woman may appeal to her insurance company by demonstrating that her symptoms require the expertise of one of these three physicians. I cannot stress how difficult these conversations are for these women, as they must provide quite intimate details in order to convince the (usually male) insurance adjuster that their condition cannot be treated adequately by a nonspecialist.
4. Out-of-state patients were instructed to perform their own research, as the list was state specific.
5. Segal is quoting Lacan's essay "Encore" (1972–1973) in her first sentence. See Mitchell and Rose (1985, 145), where the essay appears as "God and the Jouissance of the Women," for the full citation.
6. Vulvar Intraepithelial Neoplasia (precancer), at the first level of abnormality. This disease condition, because of its malignant nature and potential, would typically be managed within oncology or gynecology and outside the

confines of the VHC. Unfortunately for Colleen, her clinical presentation blurred a lot of lines.
7. To be very clear, these apologies were cut short by the physicians and other VHC staff.
8. Jim was one of only two (male) partners present for an interview. I had decided that I would let the women decide about the role they wanted their partners to play in this facet of our work together. I was not opposed to the presence of a partner, but since my primary interest was in the women's narratives (and since I harbored a mild concern that their "voices" might be muffled or diverted by the presence of a partner), I kept quiet about the subject and let my informants ask me if it was okay for him to be there. I agreed happily each time and, indeed, would welcome the opportunity to collect more of their stories, particularly since patients indicated that their husbands/partners weren't talking to *anyone*.
9. Johnson is partially quoting Dworkin (1987, 72).

CHAPTER 6. GENERATION

1. See Kurtz and Prestera (1976) for a fascinating account of a woman who presented to a bodyworker with extremely tight shoulders. When the provider began working out the tension, which involved asking and helping the patient to lower and drop her shoulders and arms, the woman began crying and resisting. Further exploration "revealed" that, as a toddler, the woman's wrists had been tied to the slats of her bed by a parent in an effort to keep the girl from reaching and touching her genitalia.
2. Libby had three sessions with Cathy prior to her vestibulectomy. Although all providers understood PT to be an optimal therapeutic strategy, Libby was the only patient with whom I worked who began it before she had her surgery. At the time of our interview, she imagined that she and David would work with a sex therapist, but neither she nor I knew how much "sex therapy" she would end up doing with Cathy.

3. This dilemma is intriguing in the context of recent research about young women's sexuality. Armstrong, Hamilton, and England (2010) found that women were more likely to orgasm during partnered sexual activity when they were in relationships (vs. casual hookups); this was the case, however, only if the couple were engaging in multiple forms of sexual activity, including cunnilingus. The forms of sex in which women were least likely to experience orgasm—casual hookups—typically included only penetrative intercourse. It is interesting to think about the generational specificity of data like these, given that it is Sharon and Anharrad's—two women in their late fifties—relationships that are the obstacle to their increased orgasmic potential.
4. Clair's surgery was planned at the time of our interview, but she had not scheduled it by the time that I departed several months later. When I last spoke to her about it, she told me that she was waiting for her husband to "lose weight," as she did not think it fair for her to "go through all that" if he "wasn't going to do anything" to make himself more attractive to her.
5. For a start, interested readers should see Cacchioni (2007), Canner (2009), Fishman (2013), New View (2000), and Tiefer (2001, 2003a, and 2003b).

CHAPTER 7. EVALUATION

1. Kempner (2014, 132–33) describes the "double jeopardy" of migraine patients whose pain no longer excuses them from responsibility, largely due to pharmaceutical ad campaigns that have successfully convinced the general public of their drug's efficacy.

EPILOGUE

1. A division of the National Institutes of Health (NIH).
2. I did not attend the 2011 meeting; the information in this chapter is gathered primarily from the resulting report (NIH 2012).

3. Botox (botulinum toxin A) has been used with some success in treating vulvar pain conditions, particularly VVS/provoked vestibulodynia. See De Andres et al. (20150; Gunter, Brewer, and Tawfik (2004); Yoon, Chung, and Shim (2006); and Petersen et al. (2009).
4. See Kempner (2014), especially pages 52–70, for an excellent discussion of how this distancing—what she calls "neuroreduction" (53)—manifests in the treatment of migraine pain.

REFERENCES

ACOG (American Congress of Obstetricians and Gynecologists). 2007. "ACOG Advises against Cosmetic Vaginal Procedures Due to Lack of Safety and Efficacy Data (Committee Opinion #378)." *Obstetrics and Gynecology* 110:737–38.

Ahmed, Sara. 2007. "A Phenomenology of Whiteness." *Feminist Theory*, 8(2): 149-68.

Ahmed, Sara. 2014. "Feminist Hurt/Feminism Hurts." *Feminist Killjoys*, July 21. Accessed July 26, 2014. http://feministkilljoys.com/2014/07/21/feminist-hurtfeminism-hurts/.

Ahmed, S. A., B. D. Hissong, D. Verthelyi, K. Donner, K. Becker, and E. Karpuzoglu-Sahin. 1999. "Gender and Risk of Autoimmune Diseases: Possible Role of Estrogenic Compounds." *Environmental Health Perspectives* 107 (5): 681–86.

Aikens, James A., Barbara Reed, Daniel Gorenflo, and Hope Haefner. 2003. "Depressive Symptoms among Women with Vulvar Dysesthesia." *American Journal of Obstetrics and Gynecology* 189 (2): 462–66.

Anderson, Warwick. 2003. *The Cultivation of Whiteness: Science, Health, and Racial Destiny in Australia*. New York: Basic Books.

Andrews, Jeffrey C. 2011. "Vulvodynia Interventions: Systematic Review and Evidence Grading." *Obstetrical and Gynecological Survey* 66 (5): 299–315.

Andrews, William C. 2000. "Approaches to Taking a Sexual History." *Journal of Women's Health and Gender-Based Medicine* 9 (Suppl. 1): S21–S24.

Angier, Natalie. 2000. *Woman: An Intimate Geography.* New York: Anchor Books.

Apte, Gail, Patricia Nelson, Jean-Michel Brismée, Gregory Dedrick, Rafael Justiz III, and Phillip Sizer. 2011. "Chronic Female Pelvic Pain—Part 1: Clinical Pathoanatomy and Examination of the Pelvic Region." *PAIN Practice* 12 (2): 88–110.

Armstrong, Elizabeth, Laura Hamilton, and Paula England. 2010. "Is Hooking Up Bad for Young Women?" *Contexts* 9 (3): 22–27.

Armstrong, Jennifer Keishin. 2010. "Our Poor Vaginas." *Sexy Feminist*, October 19. Accessed August 20, 2013. http://sexyfeminist.com/2010/10/19/our-poor-vaginas/.

Arnold, Lauren, Gloria A. Bachmann, Raymond Rosen, Sarah Kelly, and George G. Rhoads. 2006. "Vulvodynia: Characteristics and Associations with Comorbidities and Quality of Life." *Obstetrics and Gynecology* 107 (3): 617–24.

Bachmann, G., R. Rosen, V. W. Pinn, W. H. Utian, C. Ayers, R. Basson, Y. M. Binik, C. Brown, D. C. Foster, J. M. Gibbons Jr., I. Goldstein, A. Graziottin, H. K. Haefner, B. L. Harlow, S. K. Spadt, S. R. Leiblum, R. M. Masheb, B. D. Reed, J. D. Sobel, C. Veasley, U. Wesselmann, and S. S. Witkin. 2006. "Vulvodynia: A State-of-the-Art Consensus on Definitions, Diagnosis, and Management." *Journal of Reproductive Medicine* 51 (6): 447–56.

Baggish, Michael S., and John R. Miklos. 1995. "Vulvar Pain Syndrome: A Review." *Obstetrical and Gynecological Survey* 50 (8): 618–27.

Baggish, Michael S., Eddie H. M. Sze, and Robert Johnson. 1997. "Urinary Oxalate Excretion and Its Role in Vulvar Pain Syndrome." *American Journal of Obstetrics and Gynecology* 177 (3): 507–11.

Bailey, Cathryn. 1997. "Making Waves and Drawing Lines: The Politics of Defining the Vicissitudes of Feminism." *Hypatia* 12 (3): 17–28.

Balsamo, Anne. 1999. "Public Pregnancies and Cultural

Narratives of Surveillance." In *Revisioning Women, Health, and Healing: Feminist, Cultural, and Technoscience Perspectives*, edited by Adele Clarke and Virginia Olesen, 231–53. New York: Routledge.

Barad, Karen. 2007. *Meeting the Universe Halfway: Quantum Physics and the Entanglement of Matter and Meaning*. Durham: Duke University Press.

Bartky, Sandra. 1990. *Femininity and Domination: Studies in the Phenomenology of Oppression*. New York: Routledge.

Beasley, Chris. 2010. "The Elephant in the Room: Heterosexuality in Critical Gender/Sexuality Studies." *NORA—Nordic Journal of Feminist and Gender Research* 18 (3): 204–9.

———. 2011. "Libidinous Politics: Heterosex, 'Transgression' and Social Change." *Australian Feminist Studies* 26 (67): 25–40.

Beasley, Chris, Heather Brook, and Mary Holmes. 2012. *Heterosexuality in Theory and Practice*. New York: Routledge.

Begos, Kevin, Danielle Deaver, John Railey, Scott Sexton, and Paul Lombardo. 2012. *Against Their Will: North Carolina's Sterilization Program and the Campaign for Reparations*. Apalachicola: Gray Oak Books.

Bell, Kristin. 2005. "Genital Cutting and Western Discourses on Sexuality." *Medical Anthropology Quarterly* 19 (2): 125–48.

Bergeron, Sophie, Yitzchak Binik, Samir Khalife, and K. Pagidas. 1997. "Vulvar Vestibulitis Syndrome: A Critical Review." *Clinical Journal of Pain* 13 (1): 27–42.

Bergeron, Sophie, Yitzchak Binik, Samir Khalife, K. Pagidas, Howard Glazer, Marta Meana, and R. Amsel. 2001. "A Randomized Comparison of Group Cognitive-Behavioral Therapy, Surface Electromyographic Biofeedback, and Vestibulectomy in the Treatment of Dyspareunia Resulting from Vulvar Vestibulitis." *Pain* 91: 297–306.

Bergeron, Sophie, C. Bouchard, M. Fortier, Yitzchak Binik, and Samir Khalife. 1997. "The Surgical Treatment of Vulvar Vestibulitis Syndrome: A Follow-Up Study." *Journal of Sex and Marital Therapy* 23 (4): 317–25.

Bergeron, Sophie, Claudia Brown, Marie-Josee Lord, Monica Oala, Yitzchak Binik, and Samir Khalife. 2002. "Physical

Therapy for Vulvar Vestibulitis Syndrome: A Retrospective Study." *Journal of Sex and Marital Therapy* 28 (3): 183–92.

Bergeron Sophie, Samir Khalife, Howard Glazer, and Yitzchak Binik. 2008. "Surgical and Behavioral Treatments for Vestibulodynia: Two-and-One-Half Year Follow-Up and Predictors of Outcome." *Obstetrics and Gynecology* 111 (1): 159–66.

Berghmans, L. C. M., H .J. M. Hendriks, R. A. De Bie, E. S. C. Van Waalwijk, W. Van Doorn, K. Bo, and P. Van Kerrebroeck. 2000. "Conservative Treatment of Urge Urinary Incontinence in Women: A Systematic Review of Randomized Clinical Trials." *British Journal of Urology International* 85 (3): 254–63.

Berlucchi, Giovanni, and Salvatore Aglioti. 1997. "The Body in the Brain: Neural Bases of Corporeal Awareness." *Trends in Neuroscience* 20 (12): 560–64.

Berman, Jennifer, Laura Berman, and Elisabeth Bumiller. 2005. *For Women Only, Revised Edition: A Revolutionary Guide to Reclaiming Your Sex Life*. New York: Holt.

Berman, Laura, Jennifer Berman, Stan Felder, Dan Pollets, Sachin Chhabra, Marie Miles, and Jennifer Ann Powell. 2003. "Seeking Help for Sexual Function Complaints: What Gynecologists Need to Know about the Female Patient's Experience." *Fertility and Sterility* 79 (3): 572–76.

Bernheim, Charles, and Claire Kahane. 1985. *In Dora's Case: Freud-Hysteria-Feminism*. New York: Columbia University Press.

Binik, Yitzchak. 2003. "Dyspareunia Resulting from Vulvar Vestibulitis Syndrome: A Neglected Health Problem." *Psychiatric Times* 1: 69–73.

Blackman, Lisa. 2010. "Bodily Integrity." *Body and Society* 16 (3): 1–9.

Boardman, L. A., A. S. Cooper, L. R. Blais, and C. A. Raker. 2008. "Topical Gabapentin in the Treatment of Localized and Generalized Vulvodynia." *Obstetrics and Gynecology* 112 (3): 579–85.

Bodden-Heidrich, R., V. Kuppers, M. W. Beckmann, M. H. Ozornek, I. Rechenberger, and H. G. Bender. 1999.

"Psychosomatic Aspects of Vulvodynia: Comparison with the Chronic Pelvic Pain Syndrome." *Journal of Reproductive Medicine* 44 (5): 411–16.

Boellstorff, Tom. 2010. *Coming of Age in Second Life: An Anthropologist Explores the Virtually Human.* Princeton: Princeton University Press.

Bordo, Susan. 1993. *Unbearable Weight: Feminism, Western Culture, and the Body.* Berkeley: University of California Press.

Bornstein, Jacob, D. Zarfati, N. Goldshmid, Z. Stolar, and N. Lahat. 1997. "Vestibulodynia—A Subset of Vulvar Vestibulitis or a Novel Syndrome?" *American Journal of Obstetrics and Gynecology* 177 (6): 1439–43.

Boston Women's Health Book Collective. 1992. *The New Our Bodies, Ourselves.* New York: Simon and Schuster.

Bourdieu, Pierre. 1977. *Outline of a Theory of Practice.* Cambridge: Cambridge University Press.

———. 1984. *Distinction: A Social Critique of the Judgment of Taste.* Cambridge: Harvard University Press.

———. 1993. *The Field of Cultural Production.* New York: Columbia University Press.

Braidotti, Rosi. 2002. *Metamorphoses: Towards a Materialist Theory of Becoming.* Malden: Blackwell.

Braun, Virginia. 2005. "In Search of (Better) Sexual Pleasure: Female Genital 'Cosmetic' Surgery." *Sexualities* 8 (4): 407–24.

———. 2010. "Female Genital Cosmetic Surgery: A Critical Review of Current Knowledge and Contemporary Debates." *Journal of Women's Health* 19 (7): 1393–1407.

Braun, Virginia, and Celia Kitzinger. 2001. "'Snatch,' 'Hole,' or 'Honey-Pot'? Semantic Categories and the Problem of Nonspecificity in Female Genital Slang." *Journal of Sex Research* 38 (2): 146–58.

Braun, Virginia, and Leonore Tiefer. 2010. "The 'Designer Vagina' and the Pathologisation of Female Genital Diversity: Interventions for Change." *Radical Psychology* 18 (1). Accessed August 18, 2013. http://www.radicalpsychology.org/vol8-1/brauntiefer.html.

Briggs, Laura. 2000. "The Race of Hysteria: 'Overcivilization' and the 'Savage' Woman in Late-Nineteenth-Century Obstetrics and Gynecology." *American Quarterly* 52 (2): 246–73.

Brokenshire, Christopher, Ross Pagano, and James Scurry. 2014. "The Value of Histology in Predicting the Effectiveness of Vulvar Vestibulectomy in Provoked Vestibulodynia." *Journal of Lower Genital Tract Disease* 18 (2): 109–14.

Brown, Laura. 1995. "Not Outside the Range: One Feminist Perspective on Psychic Trauma." In *Trauma: Explorations in Memory*, edited by Cathy Caruth, 100–12. Baltimore: Johns Hopkins University Press.

Brownmiller, Susan. 1976. *Against Our Will: Men, Women and Rape*. New York: Bantam.

Buchan, Ann, Pat Munday, Gill Ravenhill, Annie Wiggs, and Fiona Brooks. 2007a. "A Qualitative Study of Women with Vulvodynia: I. The Journey into Treatment." *Journal of Reproductive Medicine* 52 (1): 15–18.

Bush, Judith. 2000. "'It's Just Part of Being a Woman': Cervical Screening, the Body, and Femininity." *Social Science and Medicine* 50 (3): 429–44.

Butler, Judith. 1993. *Bodies That Matter: On the Discursive Limits of Sex*. New York: Routledge.

Byrd, Julie A., Mark D. P. Davis, and Roy S. Rogers. 2004. "Recalcitrant Symptomatic Vulvar Lichen Planus." *Archives of Dermatology* 140 (6): 715–20.

Cacchioni, Thea. 2007. "Heterosexuality and the 'Labour of Love': A Contribution to Recent Debates on Female Sexual Dysfunction." *Sexualities* 10 (3): 299–320.

———. 2015. *Big Pharma, Women, and the Labour of Love*. Toronto: University of Toronto Press.

Cacchioni, Thea, and Carol Wolkowitz. 2011. "Treating Women's Sexual Difficulties: The Body Work of Sexual Therapy." *Sociology of Health and Illness* 33(2): 266–79.

Calasanti, Toni. 2007. "Bodacious Berry, Potency Wood, and the Aging Monster: Gender and Age Relations in Anti-Aging Ads." *Social Forces* 86 (1): 335–55.

Cameron, Deborah, and Don Kulick. 2003. *Language and Sexuality*. Cambridge: Cambridge University Press.

Canner, Liz. 2009. *Orgasm, Inc.* New York: First Run Features.
Carey, Benedict. 2007. "Who's Minding the Mind?" *New York Times*, July 31.
Carpenter, Laura. 2005. *Virginity Lost: An Intimate Portrait of First Sexual Experiences.* New York: New York University Press.
Carrillo Rowe, Aimee. 2008. *Power Lines: On the Subject of Feminist Alliances.* Durham: Duke University Press.
Carruthers, Glenn. 2008. "Types of Body Representation and the Sense of Embodiment." *Consciousness and Cognition* 17(4): 1302–16.
Cartwright, Lisa, and Brian Goldfarb. 2006. "On the Subject of Neural and Sensory Prostheses." In *The Prosthetic Impulse: From Posthuman Present to a Biocultural Future*, edited by Marquard Smith and Joanne Morra, 125–54. Cambridge: MIT Press.
Centers for Disease Control (CDC). 2010. "National Intimate Partner and Sexual Violence Survey (NISVS): An Overview of 2010 Summary Report Findings." Accessed August 19, 2013. http://www.cdc.gov/violenceprevention/pdf/cdc_nisvs_overview_insert_final-a.pdf.
Cerankowski, Karli June, and Megan Milks, eds. 2014. *Asexualities: Feminist and Queer Perspectives.* New York: Routledge.
Chadha, S., W. L. Gianotten, A. C. Drogendijk, W. C. Weijmar Schultz, L. A. Blindeman, and W. I. van der Meijden. 1998. "Histopathologic Features of Vulvar Vestibulitis." *International Journal of Gynecological Pathology* 17 (1): 7–11.
Chisholm, Dianne. 1994. "Irigaray's Hysteria." In *Engaging with Irigaray: Feminist Philosophy and Modern European Thought*, edited by Carolyn Burke, Naomi Schor, and Margaret Whitford, 263–84. New York: Columbia University Press.
Citterio, F. 2001. "Steroid Side Effects and Their Impact on Transplantation Outcome." *Transplantation* 72 (12 Suppl.): S75–S80.
Cixous, Hélène, and Catherine Clément. 1986. *The Newly Born Woman.* Minneapolis: University of Minnesota Press.

Clark-Flory, Tracy. 2014. "Women Obsess over the Size of Their Privates, Too." *Salon*, May 24. Accessed November 21, 2014. http://www.salon.com/2014/05/25/women_obsess_over_the_size_of_their_privates_too/.

Clarke, Adele, and Theresa Montini. 1993. "The Many Faces of RU-486: Tales of Situated Knowledges and Technological Contestations." *Science, Technology, and Human Values* 18 (1): 42–78.

Clarke, Adele, and Virginia Olesen. 1999. "Revising, Diffracting, Acting." In *Revisioning Women, Health, and Healing: Feminist, Cultural, and Technoscience Perspectives*, edited by Adele Clarke and Virginia Olesen, 3–48. New York: Routledge.

Clawson, Mary Ann. 2008. "Looking for Feminism: Racial Dynamics and Generational Investments in the Second Wave." *Feminist Studies* 34 (3): 526–54.

Colebrook, Claire. 2000. "From Radical Representations to Corporeal Becomings: The Feminist Philosophy of Lloyd, Grosz, and Gatens." *Hypatia* 15 (2): 76–93.

Connor, Jennifer J., Cassandra M. Brix, and Stephanie Trudeau-Hern. 2013. "The Diagnosis of Provoked Vestibulodynia: Steps and Roadblocks in a Long Journey." *Sexual and Relationship Therapy* 28 (4): 324–35.

Connor, Jennifer J., Bean Robinson, and Elizabeth Wieling. 2008. "Vulvar Pain: A Phenomenological Study of Couples in Search of Effective Diagnosis and Treatment." *Family Process* 47 (2): 139–55.

Coole, Diana, and Samantha Frost. 2010. "Introducing the New Materialisms." In *New Materialisms: Ontology, Agency, and Politics*, edited by Diana Coole and Samantha Frost, 1–46. Durham: Duke University Press.

Cooper, Richard A. 2003. "Medical Schools and Their Applicants: An Analysis." *Health Affairs* 22 (4): 71–84.

Cooper, S. M., X. Gao, J. Powell, and F. Wojnarowska. 2004. "Does Treatment of Vulvar Lichen Sclerosus Influence Its Prognosis?" *Archives of Dermatology* 140 (6): 702–6.

Cowan, M. 1964. "Neurohistological Changes in Lichen Simplex Chronicus." *Archives of Dermatology* 89: 562–68.

Crenshaw, Kimberlé. 1991. "Mapping the Margins: Intersectionality, Identity Politics, and Violence against Women of Color." *Stanford Law Review* 43: 1241–65.

Csordas, Thomas J. 1993. "Somatic Modes of Attention." *Cultural Anthropology* 8 (2): 135–56.

———. 1994. *The Sacred Self: A Cultural Phenomenology of Charismatic Healing*. Berkeley: University of California Press.

Cvetkovich, Ann. 2003. *An Archive of Feelings: Trauma, Sexuality, and Lesbian Public Cultures*. Durham: Duke University Press.

Dalton, Vanessa K., Hope K. Haefner, Barbara D. Reed, Sangeeta Senapati, and Ann Cook. 2002. "Victimization in Patients with Vulvar Dysesthesia/Vestibulodynia: Is There an Increased Prevalence?" *Journal of Reproductive Medicine* 47: 829–34.

Danby, Claire S., and Lynette J. Margesson. 2010. "Approach to the Diagnosis and Treatment of Vulvar Pain." *Dermatologic Therapy* 23 (5): 485–504.

Davis, Simone Weil. 2002. "Loose Lips Sink Ships." *Feminist Studies* 28 (1): 7–35.

D'Cruz, D. 2007. "Systemic Lupus Erythematosus." *Lancet* 359 (9561): 587–96.

De Andres, Jose, Nerea Sanchis-Lopez, Juan Marcos Asensio-Samper, Gustavo Fabregat-Cid, Vicente L. Villanueva-Perez, Vicente Monsalve Dolz, and Ana Minguez. 2015. "Vulvodynia: An Evidence-Based Literature Review and Proposed Treatment Algorithm." *Pain Practice* January 12: 1–33. Early online preview; accessed February 15, 2015.

de Marneffe, Daphne. 1996. "Looking and Listening: The Construction of Clinical Knowledge in Charcot and Freud." In *Gender and Scientific Authority*, edited by Barbara Laslett, Sally Gregory Kohlstedt, Helen Longino, and Evelynn Hammonds, 241–81. Chicago: University of Chicago Press.

De Preester, Helen, and Manos Tsakiris. 2009. "Body-Extension versus Body-Incorporation: Is There a Need for a Body-Model?" *Phenomenology and Cognitive Science* 8: 307–19.

Devault, Marjorie. 1990. "Talking and Listening from Women's Standpoint: Feminist Strategies for Interviewing and Analysis." *Social Problems* 37 (1): 96–116.

de Zengotita, Thomas. 2005. *Mediated: How the Media Shapes Your World and the Way You Live in It.* New York: Bloomsbury.

Didi-Huberman, Georges. 2003. *Invention of Hysteria: Charcot and the Photographic Iconography of the Salpetriere.* Cambridge: MIT Press.

Domar, Alice D. 1986. "Psychological Aspects of the Pelvic Exam: Individual Needs and Physician Involvement." *Women and Health: A Multidisciplinary Journal of Women's Health Issues* 10 (4): 75–90.

Dorkenoo, Efua. 1994. *Cutting the Rose. Female Genital Mutilation: The Practice and Its Prevention.* London: Minority Rights Group.

Douglas, Mary. 1966. *Purity and Danger: An Analysis of Concepts of Pollution and Taboo.* London: Routledge.

———. 1973. *Natural Symbols: Explorations in Cosmology.* New York: Vintage Books.

Downey, Greg. 2009. "Throwing Like a Girl('s Brain)." *Neuroanthropology*, February 1. Accessed September 28, 2013. http://neuroanthropology.net/2009/02/01/throwing-like-a-girls-brain/.

Dressler, William W., Kathryn S. Oths, and Clarence C. Gravlee. 2005. "Race and Ethnicity in Public Health Research: Models to Explain Health Disparities." *Annual Review of Anthropology* 34: 231–52.

Droegenmueller, W. 1992. *Comprehensive Gynecology*, 2nd ed. St. Louis: Mosby.

Duarte-Franco, Eliane, and Eduardo Franco. 2004. "Other Gynecologic Cancers: Endometrial, Ovarian, Vulvar and Vaginal Cancer." *BMC Women's Health* 4 (Suppl. 1): S1–S14.

Duden, Barbara. 1998. *The Woman beneath the Skin: A Doctor's Patients in Eighteenth-Century Germany.* Cambridge: Harvard University Press.

Dumit, Joseph. 2012. *Drugs for Life: How Pharmaceutical*

Companies Define Our Health. Durham: Duke University Press.

Duster, Troy. 2006. "The Molecular Reinscription of Race: Unanticipated Issues in Biotechnology and Forensic Science." *Patterns of Prejudice* 40 (4/5): 427–41.

Dworkin, Andrea. 1987. *Intercourse*. New York: Free Press.

Ebert, Teresa. 1996. *Ludic Feminism and After: Postmodernism, Desire, and Labor in Late Capitalism*. Ann Arbor: University of Michigan Press.

Edelman, Lee. 2004. *No Future: Queer Theory and the Death Drive*. Durham: Duke University Press.

Edwards, Libby. 1989. "Vulvar Lichen Planus." *Archives of Dermatology* 125 (12): 1677–80.

———. 2003. "New Concepts in Vulvodynia." *American Journal of Obstetrics and Gynecology* 189 (3, Suppl.): S24–S30.

———. 2004. "Subsets of Vulvodynia: Overlapping Characteristics." *Journal of Reproductive Medicine* 49 (11): 883–87.

Edwards, Libby, M. Mason, M. Phillips, J. Norton, and M. Boyle. 1997. "Childhood Sexual and Physical Abuse: Incidence in Patients with Vulvodynia." *Journal of Reproductive Medicine* 42: 135–39.

Ehrenreich, Barbara, and Deirdre English. 1973. *Complaints and Disorders: The Sexual Politics of Sickness*. Old Westbury: Feminist Press.

———. 1978. *For Her Own Good: 150 Years of the Experts' Advice to Women*. New York: Anchor Books.

Einstein, Gillian. 2008. "From Body to Brain: Considering the Neurobiological Effects of Female Genital Cutting." *Perspectives in Biology and Medicine* 51 (1): 84–97.

Einstein, Gillian, and Margrit Shildrick. 2009. "The Postconventional Body: Retheorising Women's Health." *Social Science and Medicine* 69 (2): 293–300.

Enloe, Cynthia. 2004. "Wielding Masculinity inside Abu Ghraib: Making Feminist Sense of an American Military Scandal." *Asian Journal of Women's Studies* 10 (3): 89–102.

Ensler, Eve. 2001. *The Vagina Monologues: The V-Day Edition*. New York: Villard.

Fabello, Melissa. 2013. "How to Start Loving Your

Vagina." *Everyday Feminism*, March 28. Accessed August 20, 2013. http://everydayfeminism.com/2013/03/how-to-start-loving-your-vagina/.

Fahs, Breanne. 2014. "'Freedom to' and 'Freedom from': A New Vision for Sex-Positive Politics." *Sexualities* 17 (3): 267–90.

Fausto-Sterling, Anne. 1995. "Gender, Race, and Nation: The Comparative Anatomy of 'Hottentot' Women in Europe, 1815–1817." In *Deviant Bodies: Critical Perspectives on Difference in Science and Popular Culture*, edited by Jennifer Terry and Jacqueline Urla, 19–48. Bloomington: Indiana University Press.

———. 2000. *Sexing the Body: Gender Politics and the Construction of Sexuality*. New York: Basic Books.

———. 2004. "Refashioning Race: DNA and the Politics of Health Care." *differences: A Journal of Feminist Cultural Studies* 15 (3): 1–37.

———. 2005. "The Bare Bones of Sex: Part 1—Sex and Gender." *Signs: Journal of Women in Culture and Society* 30 (2): 1491–1527.

Federation of Feminist Women's Health Centers. 1981. *A New View of a Woman's Body: A Fully Illustrated Guide*. New York: Simon and Schuster.

Feinstein, A. R., and R. I. Horwitz. 1997. "Problems in the 'Evidence' of 'Evidence-Based Medicine.'" *Journal of the American Medical Association* 103 (6): 529–35.

Feldman, J. L. 2008. "Medical Management of Transgender Patients." In *The Fenway Guide to Enhancing Healthcare in Lesbian, Gay, Bisexual and Transgendered Communities*, edited by H. Makadon, K. Mayer, J. Potter, and H. Goldhammer. Philadelphia: American College of Physicians.

Findlen, Barbara, ed. 2001. *Listen Up: Voices from the Next Feminist Generation*. Seattle: Seal Press.

Fisher, Jill, ed. 2011. *Gender and the Science of Difference: Cultural Politics of Contemporary Science and Medicine*. New Brunswick: Rutgers University Press.

Fisher, Sidney, and Seymour Cleveland. 1968. *Body Image and Personality*. New York: Dover.

Fishman, Jennifer. 2013. "Manufacturing Desire: The Commodification of Female Sexual Dysfunction." *Social Studies of Science* 43 (4): 187–218.

Foster, David C. 1995. "Case-Control Study of Vulvar Vestibulitis Syndrome." *Journal of Women's Health* 6: 677–80.

———. 2002. "Vulvar Disease." *Obstetrics and Gynecology* 100 (1): 145–63.

Foster, David C., Robert H. Dworkin, and Ronald W. Wood. 2005. "Effects of Intradermal Foot and Forearm Capsaicin Injections in Normal and Vulvodynia-Afflicted Women." *Pain* 117: 128–36.

Foster, David C., and J. D. Hasday. 1997. "Elevated Tissue Levels of Interleukin-1B and Tumor Necrosis Factor-a in Vulvar Vestibulitis." *Obstetrics and Gynecology* 89 (2): 291–96.

Foster, David C., Katherine H. Piekarz, Thomas I. Murant, Randi LaPoint, Constantine G. Haidaris, and Richard P. Phipps. 2007. "Enhanced Synthesis of Proinflammatory Cytokines by Vulvovestibular Fibroblasts: Implications for Vulvar Vestibulitis." *American Journal of Obstetrics and Gynecology* 196 (4): 346.e1—346.e8.

Foster, David C., Todd M. Sazenski, and Christopher J. Stodgell. 2004. "Impact of Genetic Variation in Interleukin-1 Receptor Antagonist and Melanocortin-1 Receptor Genes on Vulvar Vestibulitis Syndrome." *Journal of Reproductive Medicine* 49 (7): 503–9.

Foucault, Michel. 1973. *The Birth of the Clinic: An Archaeology of Medical Perception.* New York: Vintage Books.

———. 1990a. *The History of Sexuality.* Vol. 1, *An Introduction.* New York: Vintage.

———. 1990b. *The History of Sexuality.* Vol. 2, *The Use of Pleasure.* New York: Vintage.

Frank, Ellen. 2000. *Gender and Its Effects on Psychopathology.* Arlington: American Psychiatric Press.

Freeman, Elizabeth. 2005. "Time Binds, or, Erotohistoriography." *Social Text* 23 (3–4): 57–68.

———. 2007. "Introduction." *GLQ: A Journal of Lesbian and Gay Studies* 13 (2–3): 159–76.

French, Lindsay. 1994. "The Political Economy of Injury and

Compassion: Amputees on the Thai-Cambodia Border." In *Embodiment and Experience: The Existential Ground of Culture and Self*, edited by Thomas J. Csordas, 69–99. Cambridge: Cambridge University Press.

Freud, Sigmund. 1917a (1966). "Fixation to Traumas—The Unconscious (Lecture 18)." In *Sigmund Freud: Introductory Lectures on Psycho-Analysis (Standard Edition)*, edited by James Strachey, 338–53. New York: Norton.

———. 1917b (1958). "One of the Difficulties of Psycho-Analysis." In *On Creativity and the Unconscious: Papers on the Psychology of Art, Literature, Love, Religion*, edited by Benjamin Nelson, 1–10. New York: Harper and Brothers.

———. 1919 (1958). "The Uncanny." In *On Creativity and the Unconscious: Papers on the Psychology of Art, Literature, Love, Religion*, edited by Benjamin Nelson, 122–61. New York: Harper and Brothers.

———. 1925 (1977). "Some Psychological Consequences of the Anatomical Distinction between the Sexes." In *On Sexuality: Three Essays on the Theory of Sexuality and Other Works*, vol. 7 of The Pelican Freud Library, translated by James Strachey, edited by Angela Richards, 331–43. Harmondsworth: Penguin.

Friedrich, Eduard. 1983. *Vulvar Disease*. Philadelphia: W. B. Saunders.

Frueh, Joanna. 2003. "Vaginal Aesthetics." *Hypatia* 18 (4): 137–58.

Furlonge, C. B., R. N. Thin, B. E. Evans, and P. McKee. 1991. "Vulvar Vestibulitis Syndrome: A Clinico-Pathological Study." *BJOG: An International Journal of Obstetrics and Gynaecology* 98 (7): 703–6.

Gatens, Moira. 1996. *Imaginary Bodies: Ethics, Power and Corporeality*. London: Routledge.

Gentilcore-Saulnier, Evelyne, Linda McLean, Corrie Goldfinger, Caroline F. Pukall, and Susan Chamberlain. 2010. "Pelvic Floor Muscle Assessment Outcomes in Women with and without Provoked Vestibulodynia and the Impact of a Physical Therapy Program." *Journal of Sexual Medicine* 7 (2): 1003–22.

Gerber Stefan, Ann Marie Bongiovanni, William Ledger, and

Steven Witkin. 2002. "A Deficiency in Interferon-alpha Production in Women with Vulvar Vestibulitis." *American Journal of Obstetrics and Gynecology* 186 (3): 361–64.

Gerhard, Jane. 2005. "Sex and the City: Carrie Bradshaw's Queer Postfeminism." *Feminist Media Studies* 5 (1): 37–49.

Gilman, Sander. 1985. *Difference and Pathology: Stereotypes of Sexuality, Race and Madness*. Ithaca: Cornell University Press.

Gilroy, Paul. 2000. *Against Race: Imagining Political Culture Beyond the Color Line*. Cambridge: The Belknap Press of Harvard University Press.

Glazer, Howard, and Gae Rodke. 2002. *The Vulvodynia Survival Guide: How to Overcome Painful Vaginal Symptoms and Enjoy an Active Lifestyle*. Oakland: New Harbinger.

Goetsch, Martha. 1991. "Vulvar Vestibulitis: Prevalence and Historic Features in a General Gynecologic Practice." *American Journal of Obstetrics and Gynecology* 164 (6, pt.1): 1609–16.

———. 1996. "Simplified Surgical Revision of the Vulvar Vestibule for Vulvar Vestibulitis." *American Journal of Obstetrics and Gynecology* 174 (6): 1701–05.

———. 2007. "Surgery Combined with Muscle Therapy for Dyspareunia from Vulvar Vestibulitis: An Observational Study." *Journal of Reproductive Medicine* 52 (7): 597–603.

———. 2009. "Incidence of Bartholin's Duct Occlusion after Superficial Localized Vestibulectomy." *American Journal of Obstetrics and Gynecology* 200 (6): 688.e1–688.e6.

Goldfinger, Corrie, Caroline F. Pukall, Evelyne Gentilcore-Saulnier, Linda McLean, and Susan Chamberlain. 2009. "A Prospective Study of Pelvic Floor Physical Therapy: Pain and Psychosexual Outcomes in Provoked Vestibulodynia." *Journal of Sexual Medicine* 6 (7): 1955–68.

Goldstein, Andrew, and Lara Burrows. 2008. "Vulvodynia." *Journal of Sexual Medicine* 5 (1): 5–14.

Goodman, Alan, Deborah Heath, and M. Susan Lindee, eds. 2003. *Genetic Nature/Culture: Anthropology and Science beyond the Two-Culture Divide*. Berkeley: University of California Press.

Goodman, Michael P., Otto J. Placik, Royal H. Benson III, John

R. Miklos, Robert D. Moore, Robert A. Jason, David L. Matlock, et al. 2010. "A Large Multicenter Outcome Study of Female Genital Plastic Surgery." *Journal of Sexual Medicine* 7 (4/1): 1565–77.

Gordon, Allan S., Manijeh Panahian-Jand, Fay McComb, Chiara Melegari, and Sandi Sharp. 2003. "Characteristics of Women with Vulvar Pain Disorders: Responses to a Web-Based Survey." *Journal of Sex and Marital Therapy* 29 (1): 45–58.

Gould, Stephen Jay. 1985. *The Flamingo's Smile: Reflections on Natural History*. New York: Norton.

Gravlee, Clarence. 2009. "How Race Becomes Biology: Embodiment of Social Inequality." *American Journal of Physical Anthropology* 139 (1): 47–57.

Grant, Jaime M., Lisa A. Mottet, Justin Tanis, Jack Harrison, Jody L. Herman, and Mara Keisling. 2011. *Injustice at Every Turn: A Report of the National Transgender Discrimination Survey*. Washington: National Center for Transgender Equality and National Gay and Lesbian Task Force.

Green, Fiona J. 2005. "From Clitoridectomies to 'Designer Vaginas': The Medical Construction of Heteronormative Female Bodies and Sexuality through Female Genital 'Cutting.'" *Sexualities, Evolution, and Gender* 7 (2): 153–87.

Greer, Germaine. 1972. *The Female Eunuch*. New York: Bantam Books.

Grosz, Elizabeth. 1994. *Volatile Bodies: Towards a Corporeal Feminism*. Bloomington: Indiana University Press.

———. 1995. *Space, Time, and Perversion: Essays on the Politics of Bodies*. New York: Routledge.

Gunter, Jennifer, Alan Brewer, and Ossama Tawfik. 2004. "Botulinum Toxin A for Vulvodynia." *Journal of Pain* 5 (4): 238–40.

Haefner, Hope K., Michael E. Collins, Gordon Davis, Libby Edwards, David C. Foster, Elizabeth (Dee) Heaton Hartmann, Raymond Kaufman, Peter J. Lynch, Lynette Margesson, Micheline Moyal-Barracco, Claudia K. Piper, Barbara D. Reed, Elizabeth G. Stewart, and Edward Wilkinson. 2005. "The Vulvodynia Guideline." *Journal of Lower Genital Tract Disease* 9 (1): 40–51.

Haggerty, Catherine L., Jeffrey F. Peipert, Sherry Weitzen, Susan L. Hendrix, Robert L. Holley, Deborah B. Nelson, Hugh Randall, et al. 2005. "Predictors of Chronic Pelvic Pain in an Urban Population of Women with Symptoms and Signs of Pelvic Inflammatory Disease." *Sexually Transmitted Diseases* 32(5): 293–99.
Halberstam, Judith. 2005. *In a Queer Time and Place: Transgender Bodies, Subcultural Lives.* New York: New York University Press.
Haraway, Donna. 1989. *Primate Visions: Gender, Race, and Nature in the World of Modern Science.* New York: Routledge.
———. 1997. *Modest Witness@Second_Millennium. FemaleMan_Meets_Oncomouse: Feminism and Technoscience.* New York: Routledge.
Harding, Susan. 2000. *The Book of Jerry Falwell: Fundamentalist Language and Politics.* Princeton: Princeton University Press.
Harlow, Bernard, and Elizabeth Gunther Stewart. 2003. "A Population-Based Assessment of Chronic Unexplained Vulvar Pain: Have We Underestimated the Prevalence of Vulvodynia?" *Journal of the American Medical Women's Association* 58 (2): 82–88.
———. 2005. "Adult-Onset Vulvodynia in Relation to Childhood Violence Victimization." *American Journal of Epidemiology* 161 (9): 871–80.
Harlow Bernard L., G. Vazquez, R. F. MacLehose, D. J. Erickson, J. M. Oakes, and S. J. Duval. 2009. "Self-Reported Vulvar Pain Characteristics and Their Association with Clinically Confirmed Vestibulodynia." *Journal of Women's Health* 18: 1333–40.
Harlow, Bernard, Lauren A. Wise, and Elizabeth Stewart. 2001. "Prevalence and Predictors of Chronic Genital Discomfort." *American Journal of Obstetrics and Gynecology* 185 (3): 545–50.
Hartmann, Dee. 2010. "Chronic Vulvar Pain from a Physical Therapy Perspective." *Dermatologic Therapy* 23 (5): 505–13.
Hebdige, Dick. 1979. *Subculture: The Meaning of Style.* London: Routledge.

Henslin, James M., and Mae A. Biggs. 1971. "Dramaturgical Desexualization: The Sociology of the Vaginal Examination." In *Studies of the Sociology of Sex*, edited by James M. Henslin, 243–72. New York: Appleton-Century Crofts.

Herman, Judith. 1992. *Trauma and Recovery*. New York: Basic Books.

Hess, Amanda. 2014. "Why Women Aren't Welcome on the Internet." *Pacific Standard*, January 6. Accessed November 21, 2014. http://www.psmag.com/navigation/health-and-behavior/women-arent-welcome-internet-72170/.

Hlavka, Heather R. 2014. "Normalizing Sexual Violence: Young Women Account for Harassment and Abuse." *Gender and Society* 28 (3): 337–58.

Holland, Janet, Caroline Ramazanoglu, Sue Sharpe, and Rachel Thomson. 1998. *The Male in the Head: Young People, Heterosexuality, and Power*. London: Tufnell Press.

Horn, David G. 2003. *The Criminal Body: Lombroso and the Anatomy of Deviance*. New York: Routledge.

Hosken, Fran P. 1994. *The Hosken Report: Genital and Sexual Mutilation of Females*. Lexington: Women's International Network Press.

Houppert, Karen. 2007. "Final Period." *New York Times*, July 17.

Irigaray, Luce. 1985a. *The Speculum of the Other Woman*. Ithaca: Cornell University Press.

———. 1985b. *This Sex Which Is Not One*. Ithaca: Cornell University Press.

———. 1992. *Elemental Passions*. New York: Routledge.

———. 1993a. *Sexes and Genealogies*. New York: Columbia University Press.

———. 1993b. *The Forgetting of Air in Martin Heidegger*. Austin: University of Texas Press.

———. 2004. *Key Writings*. London: Continuum.

Jackson, Jean. 1994. "Chronic Pain and the Tension between the Body as Subject and Object." In *Embodiment and Experience: The Existential Ground of Culture and Self*, edited by Thomas J. Csordas, 201–28. Cambridge: Cambridge University Press.

Jackson, Margaret. 1987. "'Facts of Life' or the Eroticization of Women's Oppression? Sexology and the Social Construction of Heterosexuality." In *The Cultural Construction of Sexuality*, edited by Pat Caplan, 52–81. London: Tavistock.

Jackson, Stevi. 2006. "Gender, Sexuality, and Heterosexuality: The Complexity (and Limits) of Heteronormativity." *Feminist Theory* 7 (1): 105–21.

Jacob, Krista, and Adela Licona. 2005. "Writing the Waves: A Dialogue on the Tools, Tactics, and Tensions of Feminisms and Feminist Practices over Time and Place." *NWSA Journal* 17 (1): 197–205.

Jaeschke, Roman, Gordon Guyatt, David L. Sackett, Eric Bass, Patrick Brill-Edwards, George Browman, Deborah Cook, et al. (The Evidence-Based Medicine Working Group). 1994. "Users' Guides to the Medical Literature. III. How to Use an Article about a Diagnostic Test: B. What Are the Results and Will They Help Me in Caring for My Patients?" *Journal of the American Medical Association* 271 (9): 703–7.

James, Stanlie M., and Claire C. Robertson. 2005. *Genital Cutting and Transnational Sisterhood: Disputing U.S. Polemics*. Urbana: University of Illinois Press.

Jamieson, D. J., and J. F. Steege. 1996. "The Prevalence of Dysmenorrhea, Dyspareunia, Pelvic Pain, and Irritable Bowel Syndrome in Primary Care Practices." *Obstetrics and Gynecology* 87 (1): 55–58.

Jamison, Leslie. 2014. "The Empathy Exams: A Medical Actor Writes Her Own Script." *The Believer*, February. Accessed November 21, 2014. http://www.believermag.com/issues/201402/?read=article_jamison.

Jantos, M., and G. White. 1997. "The Vestibulitis Syndrome: Medical and Psychosexual Assessment of a Cohort of Patients." *Journal of Reproductive Medicine* 42 (3): 145–52.

Jarrell, John F., George A. Vilos, Catherine Allaire, Susan Burgess, and Claude Fortin. 2005. "Consensus Guidelines for the Management of Chronic Pelvic Pain." *Journal of Obstetrics and Gynaecology Canada* 27 (9): 869–910.

Jensen, Jeffrey T., Kathleen Wilder, Kirstin Carr, Jillian Romm, and Amy Hansen. 2003. "Quality of Life and Sexual Function after Evaluation and Treatment at a Referral Center for Vulvovaginal Disorders." *American Journal of Obstetrics and Gynecology* 188 (6): 1629–35.

Johnson, Merri Lee. 2002. *Jane Sexes It Up: True Confessions of Feminist Desire.* New York: Four Walls Eight Windows.

Jordan-Young, Rebecca. 2011. *Brain Storm: The Flaws in the Science of Sex Differences.* Cambridge: Harvard University Press.

Kaler, Amy. 2006. "Unreal Women: Sex, Gender, Identity, and the Lived Experience of Vulvar Pain." *Feminist Review* 82: 50–75.

Kapsalis, Terry. 1997. *Public Privates: Performing Gynecology from Both Ends of the Speculum.* Durham: Duke University Press.

Karkazis, Katrina. 2008. *Fixing Sex: Intersex, Medical Authority, and Lived Experience.* Durham: Duke University Press.

Kaysen, Susanna. 2001. *The Camera My Mother Gave Me.* New York: Alfred A. Knopf.

Kehoe, Sean, and David Luesley. 1999. "Vulvar Vestibulitis Treated by Modified Vestibulectomy." *International Journal of Gynecology and Obstetrics* 64 (2): 147–52.

Kempner, Joanna. 2014. *Not Tonight: Migraine and the Politics of Gender and Health.* Chicago: University of Chicago Press.

Kessler, Jo Marie. 1988. "When the Diagnosis Is Vaginismus: Fighting Misconceptions." *Women and Therapy* 7 (2–3): 175.

Kessler, Suzanne. 1998. *Lessons from the Intersexed.* New Brunswick: Rutgers University Press.

Kessler, Suzanne J., and Wendy McKenna. 1978. *Gender: An Ethnomethodological Approach.* New York: Wiley.

King, Samantha. 2008. *Pink Ribbons, Inc.: Breast Cancer and the Politics of Philanthropy.* Minneapolis: University of Minnesota Press.

Knoblich, Gunther, Ian Thornton, Marc Grosjean, and Maggie Sciffrar. 2006. *Human Body Perception from the Inside Out.* Oxford: Oxford University Press.

Knorr-Cetina, Karen. 1999. *Epistemic Cultures: How the Sciences Make Knowledge.* Cambridge: Harvard University Press.

Krieger, Nancy.1999. "Embodying Inequality: A Review of Concepts, Measures, and Methods for Studying Health Consequences of Discrimination" *International Journal of Health Services* 29 (2): 295–52.

———, ed. 2005. *Embodying Inequality: Epidemiologic Perspectives*. Amityville: Baywood.

Kunstfeld, Rainer, Reinhard Kirnbauer, Georg Stingl, and Franz M. Karlhofer. 2003. "Successful Treatment of Vulvar Lichen Sclerosus with Topical Tacrolimus." *Archives of Dermatology* 139 (7): 850–52.

Kurtz, Ron, and Hector Prestera. 1976. *The Body Reveals: An Illustrated Guide to the Psychology of the Body*. New York: Harper and Row.

Labuski, Christine. 2008. "Virgins at the Threshold." In *Luce Irigaray: Teaching*, edited by Luce Irigaray and Mary Green, 13–23. London: Continuum.

———. 2011. "Moving beyond the Mystery of Female Genital Pain." In *Embodied Resistance: Challenging the Norms, Breaking the Rules*, edited by Chris Bobel and Samantha Kwan, 143–55. Nashville: Vanderbilt University Press.

———. 2013. "Vulnerable Vulvas: Female Genital Integrity in Health and Disease." *Feminist Studies* 39 (1): 248–76.

———. 2014. "Deferred Desire: The Asexuality of Chronic Genital Pain." In *Asexualities: Feminist and Queer Perspectives*, edited by Karli June Cerankowski and Megan Milks, 302–28. New York: Routledge.

Landry, Tina, Sophie Bergeron, Marie-Josee Dupuis, and Genevieve Desrochers. 2008. "The Treatment of Provoked Vestibulodynia: A Critical Review." *Clinical Journal of Pain* 24 (2): 155–71.

Laqueur, Thomas W. 1990. *Making Sex: Body and Gender from the Greeks to Freud*. Cambridge: Harvard University Press.

Latour, Bruno. 2007. *Reassembling the Social: An Introduction to Actor-Network-Theory*. Oxford: Oxford University Press.

Latour, Bruno, and Steve Woolgar. 1986. *Laboratory Life: The Construction of Scientific Facts*. Princeton: Princeton University Press.

Lavy, Rebecca, Linda Hynan, and Robert W. Haley. 2007. "Prevalence of Vulvar Pain in an Urban, Minority Population." *Journal of Reproductive Medicine* 52 (1): 59–62.

Lawhead, R. A. 1990. "Vulvar Self-Examination: What Your Patients Should Know." *Physician Assistant* (November): 55–62.

Learman, Lee A. 2005. "Chronic Pelvic Pain—Part 1: Prevalence, Evaluation, Etiology, and Comorbidities." *Johns Hopkins Advanced Studies in Medicine* 5 (6): 306–15.

Leclair, Catherine, Martha Goetsch, K. K. Lee, and Jeffrey Jensen. 2007. "KTP-Nd: YAG Laser Therapy for the Treatment of Vestibulodynia: A Follow-Up Study." *Journal of Reproductive Medicine* 52 (1): 53–58.

Leclair, Catherine, and Jeffrey Jensen. 2005. "A Systematic Approach to Vulvodynia." *Current Women's Health Reviews* 1 (3): 209–16.

Leder, Drew. 1990. *The Absent Body*. Chicago: University of Chicago Press.

Lerner, Harriet. 2005. "Speaking the Unspeakable: Another Secret V Word." *Lilith* 30 (1): 28–30.

Lewine, Howard. 2014. "Expert Panel Says Healthy Women Don't Need Yearly Pelvic Exam." *Harvard Health Blog*, July 2. Accessed November 21, 2014. http://www.health.harvard.edu/blog/expert-panel-says-healthy-women-dont-need-yearly-pelvic-exam-201407027250.

Lightfoot-Klein, Hanny. 1989. *Prisoners of Ritual: An Odyssey into Female Genital Circumcision in Africa*. New York: Haworth Press.

Livingston, Julie. 2012. *Improvising Medicine: An African Oncology Ward in an Emerging Cancer Epidemic*. Durham: Duke University Press.

Lock, Margaret, Julia Freeman, Gillian Chillibeck, Briony Beveridge, and Miriam Padolsky. 2007. "Susceptibility Genes and the Question of Embodied Identity." *Medical Anthropology Quarterly* 21 (3): 256–76.

Lock, Margaret, and Patricia Kaufert. 1998. *Pragmatic Women and Body Politics*. Cambridge: Cambridge University Press.

Losin, Elizabeth. 2013. "The Making of a Cultural Neuroscientist."

Neuroanthropology, March 11. Accessed November 21, 2014. http://blogs.plos.org/neuroanthropology/2013/03/11/the-making-of-a-cultural-neuroscientist/.

Lotery, Helen, and Rudolph P. Galask. 2003. "Erosive Lichen Planus of the Vulva and Vagina." *Obstetrics and Gynecology* 101: 1121–25.

Lynch, Peter J. 1986. "Vulvodynia: A Syndrome of Unexplained Vulvar Pain, Psychologic Disability, and Sexual Dysfunction." *Journal of Reproductive Medicine* 31: 773–80.

———. 2004. "Lichen Simplex Chronicus (Atopic/Neurodermatitis) of the Anogenital Region." *Dermatologic Therapy* 17 (1): 8–19.

Manderson, Lenore. 2011. *Surface Tensions: Surgery, Bodily Boundaries, and the Social Self*. Walnut Creek: Left Coast Press.

Marshall, Barbara. 2006. "The New Virility: Viagra, Male Aging, and Sexual Function." *Sexualities* 9 (3): 345–62.

Martin, Emily. 1987. *The Woman in the Body: A Cultural Analysis of Reproduction*. Boston: Beacon.

———. 1994. *Flexible Bodies: The Role of Immunity in American Culture from the Days of Polio to the Age of AIDS*. Boston: Beacon.

———. 2009. *Bipolar Expeditions: Mania and Depression in American Culture*. Princeton: Princeton University Press.

Masheb, Robin M., Elizabeth Brondolo, and Robert D. Kerns. 2002. "A Multidimensional, Case-Control Study of Women with Self-Identified Chronic Vulvar Pain." *Pain Medicine* 3 (3): 253–59.

Masheb, Robin M., Justin M. Nash, Elizabeth Brondolo, and Robert D. Kerns. 2000. "Vulvodynia: An Introduction and Critical Review of a Chronic Pain Condition." *Pain* 86 (1–2): 3–10.

Masson, Jeffrey. 2003. *The Assault on Truth: Freud's Suppression of the Seduction Theory*. New York: Ballantine.

Mayer, Horacio F., Maria Laura B. de Elizalde, Natalie Duh, and Hugo D. Loustau. 2011. "Bidimensional Labia Minora Reduction." *European Journal of Plastic Surgery* 34 (5): 345–50.

McEwan, Melissa. 2014. "We Need to Talk about This." *Shakesville*, June 10. Accessed July 26, 2014. http://www.shakesville.com/2014/06/we-need-to-talk-about-this.html.

McGarvey, E., C. Peterson, R. Pinkerton, A. Keller, and A. Clayton. 2003. "Medical Students' Perceptions of Sexual Health Issues Prior to a Curriculum Enhancement." *International Journal of Impotence Research* 15 (Suppl. 5): S58–S66.

M'Charek, Amade. 2013. "Beyond Fact or Fiction: On the Materiality of Race in Practice." *Cultural Anthropology* 28 (3): 420–42.

Mehrotra, Ateev, Alan M. Zaslavsky, and John A. Ayanian. 2007. "Preventive Health Examinations and Preventive Gynecological Examinations in the United States." *Archives of Internal Medicine* 167 (17): 1876–83.

Merleau-Ponty, Maurice. 1962. *Phenomenology of Perception*. London: Routledge.

Metzl, Jonathan M., and Helena Hansen. 2014. "Structural Competency: Theorizing a New Medical Engagement with Stigma and Inequality." *Social Science and Medicine* 103: 126–33.

Metzl, Jonathan, and Anna Kirkland, eds. 2010. *Against Health: How Health Became the New Morality*. New York: New York University Press.

Michael, Mike, and Marsha Rosengarten. 2012. "Medicine: Experimentation, Politics, Emergent Bodies." *Body and Society* 18 (3–4): 1–17.

Mitchell, Juliet, and Jacqueline Rose, eds. 1985. *Feminine Sexuality: Jacques Lacan and the École Freudienne*. New York: W.W. Norton and Company.

Mohanty, Chandra Talpade. 1991. "Under Western Eyes: Feminist Scholarship and Colonial Discourses." In *Third World Women and the Politics of Feminism*, edited by Chandra Talpade Mohanty, Ann Russo, and Lourdes Torres, 51–80. Bloomington: Indiana University Press.

Mol, Annemarie. 2002. *The Body Multiple: Ontology in Medical Practice*. Durham: Duke University Press.

Moore, Lisa Jean, and Adele Clarke. 1995. "Clitoral Conventions and Transgressions: Graphic Representations

in Anatomy Texts, c. 1900–91." *Feminist Studies* 21 (2): 255–301.
Moses, Claire Goldberg. 2012. "'What's in a Name?' On Writing the History of Feminism." *Feminist Studies* 38 (3): 757–79.
Mulla, Sameena. 2014. *The Violence of Care: Rape Victims, Forensic Nurses, and Sexual Assault Intervention.* New York: New York University Press.
Munday, Pat, Ann Buchan, Gill Ravenhill, Annie Wiggs, and Fiona Brooks. 2007. "A Qualitative Study of Women with Vulvodynia: II. Response to a Multidisciplinary Approach to Management." *Journal of Reproductive Medicine* 52 (1): 19–22.
Murphy, Marilyn. 1991. "The Power of Naming." *Off Our Backs* 21 (4): 13.
Murphy, Michelle. 2012. *Seizing the Means of Reproduction: Entanglements of Feminism, Health, and Technoscience.* Durham: Duke University Press.
Muscio, Inga. 1998. *Cunt: A Declaration of Independence.* New York: Seal Press.
National Institutes of Health (NIH), Eunice Kennedy Shriver National Institute of Child Health and Human Development (NICHD). 2012. *Research Plan on Vulvodynia.* Accessed September 22, 2013. http://www.nichd.nih.gov/publications/pubs/Documents/NIH_Vulvodynia_Plan_April2012.pdf.
National Vulvodynia Association (NVA). 2007. "NIH Launches Vulvodynia Awareness Campaign." *NVA News* 39 (Winter). Accessed August 19, 2013. http://www.nva.org/nih_awareness.html.
Neilson, Brett. 2012. "Ageing, Experience, Biopolitics: Life's Unfolding." *Body and Society* 18 (3–4): 44–71.
Nelson, Patricia, Gail Apte, Jean-Michel Brismée, Gregory Dedrick, Rafael Justiz III, and Phillip Sizer. 2011. "Chronic Female Pelvic Pain—Part 2: Differential Diagnosis and Management." *PAIN Practice* 12 (2): 111–41.
New View Campaign (The Working Group on a New View of Women's Sexual Problems). 2000. "The New View

Manifesto." Accessed August 18, 2013. http://www.newviewcampaign.org/manifesto1.asp.

Ng, C. J., and S. A. McCarthy. 2002. "Teaching Medical Students How to Take a Sexual History and Discuss Sexual Health Issues." *Medical Journal of Malaysia* 57 (Suppl. E): 44–51.

Nguyen, Ruby H. N., Robyn L. Reese, and Bernard L. Harlow. 2015. "Differences in Pain Subtypes between Hispanic and Non-Hispanic Women with Chronic Vulvar Pain." *Journal of Women's Health* 24 (2): 1–7.

Northrup, Christiane. 1994. *Women's Bodies, Women's Wisdom*. New York: Bantam.

Numakura, Kazuyuki, Shigeru Satoh, Norihiko Tsuchiya, Yohei Horikawa, Takamitsu Inoue, Hideaki Kakinuma, Shinobu Matsuura, et al. 2005. "Clinical and Genetic Risk Factors for Posttransplant Diabetes Mellitus in Adult Renal Transplant Recipients Treated with Tacrolimus." *Transplantation* 80 (10): 1419–24.

O'Brien, Tim. 1990. *The Things They Carried*. Boston: Houghton Mifflin.

Ogden, Gina. 1999. "Coming Home to Our Sexual Selves." *Journal of Sex Research* 36 (4): 413–14.

———. 2003. "Spiritual Dimensions of Sex Therapy: An Integrative Approach for Women." *Contemporary Sexuality* 37 (1): 13–19.

Olff, Miranda, Willie Langland, Nel Draijer, and Berthold Gersons. 2007. "Gender Differences in Posttraumatic Stress Disorder." *Psychological Bulletin* 133 (2): 183–204.

Omi, Michael, and Howard Winant. 1994. *Racial Formation in the United States: From the 1960s to the 1990s*. New York: Routledge.

Omole, Folashade, Barbara J. Simmons, and Yolanda Hacker. 2003. "Management of Bartholin's Duct Cyst and Gland Abscesses." *American Family Physician* 68 (1): 135–40.

The Onion. 2009. "Renowned Hoo-Ha Doctor Wins Nobel Prize for Medical Advancements Down There." March 30. Accessed August 18, 2013. http://www.theonion.com/articles/renowned-hooha-doctor-wins-nobel-prize-for-medical,2692/.

Ordover, Nancy. 2003. *American Eugenics: Race, Queer Anatomy, and the Science of Nationalism*. Minneapolis: University of Minnesota Press.

Ormerod, A. D. 2005. "Topical Tacrolimus and Pimecrolimus and the Risk of Cancer: How Much Cause for Concern?" *British Journal of Dermatology* 153 (4): 701–5.

Ostrow, D. N. 1980. "Surrogate Patients in Medical Education." *Innovations in Education and Teaching International* 17 (2): 82–89.

Pages, Ines-Helen, Silke Jahr, Michael Schaufele, and Eberhard Conradi. 2001. "Comparative Analysis of Biofeedback and Physical Therapy for Treatment of Urinary Stress Incontinence in Women." *American Journal of Physical Medicine and Rehabilitation* 80 (7): 494–502.

Payne, Kimberly A., Elke Reissing, Marie-Andree Lahaie, Yitzchak M. Binik, and Rhonda Amsel. 2005. "What Is Sexual Pain? A Critique of DSM's Classification of Dyspareunia and Vaginismus." *Journal of Psychology and Human Sexuality* 17 (3/4): 141–154.

Petersen, Christina D., Annamaria Giraldi, Lene Lundvall, and Ellids Kristensen. 2009. "Botulinum Toxin Type A—A Novel Treatment for Provoked Vestibulodynia? Results from a Randomized, Placebo Controlled, Double Blinded Study." *Journal of Sexual Medicine* 6 (9): 2523–37.

Piccinelli, Marco, and Greg Wilkinson. 2000. "Gender Differences in Depression: Critical Review." *British Journal of Psychiatry* 203 (2): 486–92.

Pitts, Marian K. 2009. "Transgender People in Australia and New Zealand: Health, Well-Being and Access to Health Services." *Feminism and Psychology* 19: 475–95.

Pollock, Anne. 2012. *Medicating Race: Heart Disease and Durable Preoccupations with Difference*. Durham: Duke University Press.

Potts, Annie. 2002. *The Science/Fiction of Sex: Feminist Deconstruction and the Vocabularies of Heterosex*. London: Routledge.

———. 2008. "The Female Sexual Dysfunction Debate: Different 'Problems,' New Drugs—More Pressures?" In *Contesting Illness: Processes and Practices*, edited by Pamela Moss and Katherine Teghtsoonian, 259–80. Toronto: University of Toronto Press.

Povinelli, Elizabeth. 2006. *The Empire of Love: Toward a Theory of Intimacy, Genealogy, and Carnality.* Durham: Duke University Press.

Prentice, Rachel. 2012. *Bodies in Formation: An Ethnography of Anatomy and Surgery Education.* Durham: Duke University Press.

Price, Janet, and Margrit Shildrick. 1999. "Openings on the Body: A Critical Introduction." In *Feminist Theory and the Body: A Reader*, edited by Janet Price and Margrit Shildrick, 1–14. Edinburgh: Edinburgh University Press.

Pucheu, S. 1998. "Joan: 'It Itches, It Burns': Psychoanalytic Approach to a Case of Vulvar Burning Syndrome." *Journal of Psychosomatic Obstetrics and Gynaecology* 19 (4): 175–81.

Rabinow, Paul. 1997. *Making PCR: A Story of Biotechnology.* Chicago: University of Chicago Press.

Rahman, Anika, and Nahid Toubia. 2000. *Female Genital Mutilation: A Practical Guide to Worldwide Laws and Policies.* London: Zed Press.

Rape, Abuse, and Incest National Network (RAINN). 2012. "Who Are the Victims?" Accessed August 19, 2013. http://www.rainn.org/get-information/statistics/sexual-assault-victims.

Raphael, Karen G., Cathy Spatz Widom, and Gudrum Lange. 2001. "Childhood Victimization and Pain in Adulthood: A Prospective Investigation." *Pain* 92 (1–2): 283–93.

Rapkin, Andrea, John S. McDonald, and Melinda Morgan. 2008. "Multilevel Local Anesthetic Nerve Blockade for the Treatment of Vulvar Vestibulitis Syndrome." *American Journal of Obstetrics and Gynecology* 198 (1): 41.e1–41.e5.

Reardon, Jenny. 2012. "The Democratic, Anti-Racist Genome? Technoscience at the Limits of Liberalism. *Science as Culture* 21 (1): 25–47.

Reed, Barbara, Amy Caron, Daniel Gorenflo, and Hope Haefner. 2006. "Treatment of Vulvodynia with Tricyclic

Antidepressants: Efficacy and Associated Factors." *Journal of Lower Tract Genital Disease* 10 (4): 245–51.

Reed, Barbara, Scott Crawford, Mick Couper, Christin Cave, and Hope Haefner. 2004. "Pain at the Vulvar Vestibule: A Web-Based Survey." *Journal of Lower Genital Tract Disease* 8 (1): 48–57.

Reed, Barbara, Hope Haefner, and Libby Edwards. 2008. "A Survey on Diagnosis and Treatment of Vulvodynia among Vulvodynia Researchers and Members of the International Society for the Study of Vulvovaginal Disease." *Journal of Reproductive Medicine* 53 (12): 921–29.

Reed, Barbara D., Hope K. Haefner, Sioban D. Harlow, D. W. Gorenflo, and Ananda Sen. 2006. "Reliability and Validity of Self-Reported Symptoms for Predicting Vulvodynia. *Obstetrics and Gynecology* 108 (4): 906–13.

Reed, Barbara, Hope Haefner, Margaret Punch, Randy Roth, Daniel Gorenflo, and Brenda Gillespie. 2000. "Psychosocial and Sexual Functioning in Women with Vulvodynia and Chronic Pelvic Pain: A Comparative Evaluation." *Journal of Reproductive Medicine* 45 (8): 624–32.

Reed, Barbara D., Sioban D. Harlow, Ananda Sen, Laurie Jo Legocki, Rayna M. Edwards, Nora Arato, and Hope K. Haefner. 2012. "Prevalence and Demographic Characteristics of Vulvodynia in a Population-Based Sample." *American Journal of Obstetrics and Gynecology* 206 (2): 170. e1–170.e9.

Reis, Elizabeth. 2012. *Bodies in Doubt: An American History of Intersex*. Baltimore: Johns Hopkins University Press.

Reissing, Elke, Yitzchak Binik, and Samir Khalife. 1999. "Does Vaginismus Exist? A Critical Review of the Literature." *Journal of Nervous and Mental Disease* 187 (5): 261–74.

Reissing, Elke D., C. Brown, M. Lord, Yitzchak Binik, and Samir Khalife. 2005. "Pelvic Floor Muscle Functioning in Women with Vulvar Vestibulitis Syndrome." *Journal of Psychosomatic Obstetrics and Gynecology* 26 (2): 107–13.

Renaud-Vilmer, Catherine, Benedicte Cavelier-Balloy, Raphael Porcher, and Louis Dubertret. 2004. "Vulvar Lichen

Sclerosus: Effect of Long-Term Topical Application of a Potent Steroid on the Course of the Disease." *Archives of Dermatology* 140 (6): 709–12.

Roberts, Christine. 2012. "Michigan State Rep. Lisa Brown Silenced after 'Vagina' Comments." *New York Daily News*, June 15. Accessed September 22, 2013. http://www.nydailynews.com/news/politics/michigan-state-rep-lisa-brown-silenced-vagina-comments-article-1.1096480.

Roberts, Dorothy. 1997. *Killing the Black Body: Race, Reproduction, and the Meaning of Liberty*. New York: Pantheon.

Root, Maria. 1992. "Reconstructing the Impact of Trauma on Personality." In *Personality and Psychopathology: Feminist Reappraisals*, edited by Laura Brown and Mary Ballou, 229–66. New York: Guilford.

Rosen, Raymond, David Kountz, Tracey Post-Zwicker, Sandra Leiblum, and Markus Wiegel. 2006. "Sexual Communication Skills in Residency Training: The Robert Wood Johnson Model." *Journal of Sexual Medicine* 3 (1): 37–46.

Ruch, Libby O., Susan Meyers Chandler, and Richard A. Harter. 1980. "Life Change and Rape Impact." *Journal of Health and Social Behavior* 21 (3): 248–60.

Rushin, Kate. 1989. "Naming Ourselves and Our Community: Two Viewpoints on the Call to Use the Term 'African American'; We Are the Proper Name—We Are the Subject." *Gay Community News* 16 (31): 11.

Russo, Mary. 1995. *The Female Grotesque: Risk, Excess and Modernity*. New York: Routledge.

Sacks, Oliver. 1987. *The Man Who Mistook His Wife for a Hat, and Other Clinical Tales*. New York: Harper and Row.

———. 1995. *An Anthropologist on Mars*. New York: Random House.

Sadownik, Leslie Ann. 1999. "Demographic Profile of Vulvodynia Patients." *NVA News* 5 (1): 1–7.

Salamon, Gayle. 2010. *Assuming a Body: Transgender and Rhetorics of Materiality*. New York: Columbia University Press.

Sanghavi, Darshak. 2009. "Cooking the Books: The Statistical Games behind 'Off-Label' Prescription Drug Use." *Slate*, December 21. Accessed August 19, 2013. http://www.

slate.com/articles/news_and_politics/prescriptions/2009/12/cooking_the_books.html.
Saunders, Barry. 2008. *CT Suite: The Work of Diagnosis in the Age of Noninvasive Cutting.* Durham: Duke University Press.
Savage, Dan. 1998. *Savage Love: Straight Answers from Americas Most Popular Sex Columnist.* New York: Penguin.
Scarry, Elaine. 1985. *The Body in Pain: The Making and Unmaking of the World.* New York: Oxford University Press.
Scheper-Hughes, Nancy, and Margaret Lock. 1987. "The Mindful Body: A Prolegomenon to Future Work in Medical Anthropology." *Medical Anthropology Quarterly* 1 (1): 6–41.
Scheurich, Neil. 2000. "Hysteria and the Medical Narrative." *Perspectives in Biology and Medicine* 43 (4): 461–76.
Schiebinger, Londa. 1993. *Nature's Body: Gender in the Making of Modern Science.* Boston: Beacon.
Schilder, Paul. 1950. *The Image and Appearance of the Human Body: Studies in the Constructive Energies of the Psyche.* New York: International Universities Press.
Schoen, Joanna. 2005. *Choice and Coercion: Birth Control, Sterilization, and Abortion in Public Health and Welfare.* Chapel Hill: University of North Carolina Press.
Schover, L. R., D. D. Youngs, and R. Cannata. 1992. "Psychosexual Aspects of the Evaluation and Management of Vulvar Vestibulitis." *American Journal of Obstetrics and Gynecology* 167 (3): 630–36.
Schulte, Stephanie Ricker. 2011. "Surfing Feminism's Online Wave: The Internet and the Future of Women." *Feminist Studies* 37 (3): 727–44.
Schutz, Alfred. 1972. *The Phenomenology of the Social World.* London: Heinemann.
Segal, Lynne. 1994. *Straight Sex: Rethinking the Politics of Pleasure.* Berkeley: University of California Press.
Shorter, Edward. 1992. *From Paralysis to Fatigue: A History of Psychosomatic Illness in the Modern Era.* New York: Free Press.
Showalter, Elaine. 1985. *The Female Malady: Women, Madness, and Culture in England, 1830–1980.* New York: Pantheon.

———. 1997. *Hystories: Hysterical Epidemics and Modern Media*. New York: Columbia University Press.

Silverstein, Brett. 2002. "Gender Differences in the Prevalence of Somatic versus Pure Depression: A Replication." *American Journal of Psychiatry* 159 (6): 1051–52.

Smedley, B., Y. Stith, and R. Nelson, eds. 2003. *Unequal Treatment: Confronting Racial and Ethnic Disparities in Health Care*. Washington, DC: National Academies Press.

Smelik, Anneke, and Nina Lykke. 2008. *Bits of Life: Feminism at the Intersections of Media, Bioscience, and Technology*. Seattle: University of Washington Press.

Smith, W., J. Betancourt, M. Wynia, J. Bussey-Jones, V. Stone, C. Philips, et al. 2007. "Recommendations for Teaching about Racial and Ethnic Disparities in Health and Health Care." *Annals of Internal Medicine* 152 (10): 654–65.

Smith, Yolanda R., and Hope K. Haefner. 2004. "Vulvar Lichen Sclerosus: Pathophysiology and Treatment." *American Journal of Clinical Dermatology* 5 (2): 105–25.

Sobchack, Vivian. 2010. "Living a 'Phantom Limb': On the Phenomenology of Bodily Integrity." *Body and Society* 16 (3): 51–67.

Solnit, Rebecca. 2008. "Men Explain Things to Me." *TomDispatch.com*. April 13. Accessed November 22, 2014. http://www.tomdispatch.com/post/174918.

———. 2014. "#YesAllWomen Changes the Story of the Isla Vista Massacre." *The Nation*, June 2. Accessed November 22, 2014. http://www.thenation.com/article/180077/yesallwomen-changes-story-isla-vista-massacre#.

Spade, Dean. 2006. "Mutilating Gender." In *The Transgender Studies Reader*, edited by Susan Stryker and Stephen Wittle, 315–32. New York: Routledge.

Spiegelman, Art. 2004. *In the Shadow of No Towers*. New York: Pantheon.

Springer, Kristen W., Jeanne Mager Stellman, and Rebecca Jordan-Young. 2012. "Beyond a Catalogue of Differences: A Theoretical Frame and Good Practice Guidelines for Researching Sex/Gender in Human Health." *Social Science and Medicine* 74 (11): 1817–24.

Stanbury, Rosalyn, and Elizabeth Graham. 1998. "Systemic

Corticosteroid Therapy—Side Effects and Their Management." *British Journal of Ophthalmology* 82 (6): 704–8.

Stewart, Kathleen. 2010. "Afterword: Worlding Refrains." In *The Affect Theory Reader*, edited by Melissa Gregg and Gregory J. Seigworth, 339–54. Durham: Duke University Press.

Stryker, Susan, and Stephen Whittle. 2006. *The Transgender Studies Reader*. New York: Routledge.

Szymanski, Dawn F., and Kimberly F. Balsam. 2010. "Insidious Trauma: Examining the Relationship between Heterosexism and Lesbians' PTSD Symptoms." *Traumatology* 17 (2): 4–13.

Talley, Nicholas. 2001. Serotonergic Neuroenteric Modulators. *Lancet* 358 (9298): 2061–68.

Tapper, Melbourne. 1999. *In the Blood: Sickle Cell Anemia and the Politics of Race*. Philadelphia: University of Pennsylvania Press.

Taylor, Janelle. 2005. "Surfacing the Body Interior." *Annual Review of Anthropology* 34: 741–56.

Terry, Jennifer. 1989. "The Body Invaded: Medical Surveillance of Women as Reproducers." *Socialist Review* 89 (3): 13–43.

———. 1995. "Anxious Slippages between 'Us' and 'Them': A Brief History of the Scientific Search for Homosexual Bodies." In *Deviant Bodies: Critical Perspectives on Difference in Science and Popular Culture*, edited by Jennifer Terry and Jacqueline Urla, 129–69. Bloomington: Indiana University Press.

Tiefer, Leonore. 2001. "Arriving at a 'New View' of Women's Sexual Problems: Background, Theory, and Activism." *Women and Therapy* 24 (1/2): 63–98.

———. 2003a. "The Pink Viagra Story: We Have the Drug, but What's the Disease?" *Radical Philosophy* 21 (Sept/Oct): 2–5.

———. 2003b. "Taking a Biological Turn: The Push for a 'Female Viagra' and the Medicalization of Women's Sexual Problems." *Women's Studies Quarterly* 31 (1/2): 42–54.

———. 2006. "The Viagra Phenomenon." *Sexualities* 16 (3–4): 273–94.

———. 2010. "Female Genital Cosmetic Surgery (FGCS): How the Franchise Business Met 'The Body Project.'"

Paper presented at the conference *Framing the Vulva: Genital Cosmetic Surgery and Genital Diversity*. Las Vegas, Nevada, September 26.

Tommola, Paivi, Leila Unkila-Kallio, and Jorma Paavonen. 2011. "Long-Term Follow-Up of Posterior Vestibulectomy for Treating Vulvar Vestibulitis." *Acta Obstetrica et Gynecologica Scandinavica* 90 (11): 1225–31.

Tsouroufil, Maria. 2012. "Breaking in and Breaking Out of Medical School: Feminist Academic Interrupted." *Equality, Diversity, and Inclusion: An International Journal* 31 (5/6): 467–83.

Tuma, Ruba, and Jacob Bornstein. 2006. "Vulvar Pain Syndrome (Vulvodynia)—Dilemmas in Terminology." *Harefuah* 145 (3): 215–18.

Tympanidis, P., G. Terenghi, and P. Dowd. 2002. "Increased Innervation of the Vulval Vestibule in Patients with Vulvodynia." *British Journal of Dermatology* 148 (5): 1021–27.

Valentine, David. 2007. *Imagining Transgender: An Ethnography of a Category.* Durham: Duke University Press.

———. 2012. "Sue E. Generous: Toward a Theory of Non-Transsexuality." *Feminist Studies* 38 (1): 185–211.

Valentine, David, and Riki Anne Wilchins. 1997. "One Percent on the Burn Chart: Gender, Genitals, and Hermaphrodites with Attitude." *Social Text* 52/53: 215–22.

Vance, Carole S. 1993. "Pleasure and Danger: Toward a Politics of Sexuality." In *Pleasure and Danger: Exploring Female Sexuality*, edited by Carole S. Vance, 1–28. New York: Harper Collins.

Ventolini, Gary. 2013. "Vulvar Pain: Anatomic and Recent Pathophysiologic Considerations." *Clinical Anatomy* 26 (1): 130–33.

Ventolini, Gary, and Sheela M. Barhan. 2008. "Vulvodynia." *Dermatology Online Journal* 14 (1): 2.

Virgil, Annarosa, Sandra Bacilieri, and Monica Corazza. 2001. "Managing Lichen Simplex Chronicus." *Journal of Reproductive Medicine* 46 (4): 343–46.

Weismantel, Mary. 2001. *Cholas and Pishtacos: Stories of Race and Sex in the Andes.* Chicago: University of Chicago Press.

Weiss, Gail. 1999. *Body Images: Embodiment and Intercorporeality*. London: Routledge.
Welter, Barbara. 1966. "The Cult of True Womanhood: 1820–1860." *American Quarterly* 18 (2): 151–74.
Wentzell, Emily. 2013. *Maturing Masculinities: Aging, Chronic Illness, and Viagra in Mexico*. Durham: Duke University Press.
West, Lindy. 2012. "I Don't Care about Your Stupid Vulva, It's All Vagina to Me." *Jezebel*, June 26. Accessed August 18, 2013. http://jezebel.com/5921451/i-dont-care-about-your-stupid-vulva-its-all-vagina-to-me.
Whitacre, Caroline, Stephen C. Reingold, Patricia O'Looney, Elizabeth Blankenhorn, et al. 1999. "A Gender Gap in Autoimmunity: Task Force on Gender, Multiple Sclerosis, and Autoimmunity." *Science* 283 (5406): 1277–78.
Wiley, Andrea S., and John S. Allen. 2009. *Medical Anthropology: A Biocultural Approach*. Oxford: Oxford University Press.
Williams, Raymond. 1978. *Marxism and Literature*. Oxford: Oxford University Press.
Wilson, Elizabeth. 2004. *Psychosomatic: Feminism and the Neurological Body*. Durham: Duke University Press.
Winkler, Cathy, and Kate Wininger. 1994. "Rape Trauma: Contexts of Meaning." In *Embodiment and Experience: The Existential Ground of Culture and Self*, edited by Thomas J. Csordas, 248–68. Cambridge: Cambridge University Press.
Wojnarowska, Fenella, Richard Mayou, Sue Simkin, and Alex Day. 1997. "Psychological Characteristics and Outcome of Patients Attending a Clinic for Vulval Disease." *Journal of the European Academy of Dermatology and Venereology* 8 (2): 121–29.
Wolf, Jessica. 2014. "Less Talk, More Therapy." *New York Times*, November 17. Accessed November 17, 2014. http://opinionator.blogs.nytimes.com/2014/11/17/less-talk-more-therapy/?_r=0
Wolf, Naomi. 2012. *Vagina: A New Biography*. New York: Ecco.
Wood, Ann Douglas. 1973. "'The Fashionable Diseases':

Women's Complaints and Their Treatments in Nineteenth-Century America." *Journal of Interdisciplinary History* 4 (1): 25–52.
Xu, Jian. 1999. "Body, Discourse, and the Cultural Politics of Contemporary Chinese Qigong." *Journal of Asian Studies* 58 (4): 961–91.
Yoon, H., W. S. Chung, and B. S. Shim. 2006. "Botulinum Toxin A for the Management of Vulvodynia." *International Journal of Impotence Research* 19: 84–87.
Young, Iris Marion. 2005. "Throwing Like a Girl: A Phenomenology of Feminine Body Comportment, Motility, and Spatiality." In *On Female Body Experience: "Throwing Like a Girl" and Other Essays*, 27–45. New York: Oxford University Press.
Young, Katharine. 1997. *Presence in the Flesh: The Body in Medicine*. Cambridge: Harvard University Press.
Zimmer, Carl. 2007. "In Ducks, War of the Sexes Plays Out in the Evolution of Genitalia." *New York Times*, May 1. Accessed September 16, 2013. http://www.nytimes.com/2007/05/01/science/01duck.html.
Zolnoun, Denniz, Katherine Hartmann, and John Steege. 2003. "Overnight 5% Lidocaine Ointment for Treatment of Vulvar Vestibulitis." *Obstetrics and Gynecology* 102 (1): 84–87.

INDEX

absence, xv, 3, 18, 19, 31, 32, 100, 158, 180–81, 182, 184–88; and (hyper)presence, xv, 29–32, 89, 120–21, 184–85, 240; as accumulation, 79–81, 120–21, 204, 209; as censorship and erasure, 93–96, 127, 150–51, 177–78. *See also* invisibility

Absent Body, The (Drew Leder), 182

abuse, sexual, 24–27, 98, 107, 267n5; relationship to genital pain, 25–26, 124–29, 132, 159–60, 247–48, 253. *See also* unwanted genital experience

access, 60–61, 67, 116, 193, 219, 245–47; to alternative imaginaries, 115–16, 208–09, 218–20; to the body/vulva, 2–3, 14, 88–89, 141, 158, 168, 170–73, 197–98, 203–05, 220; to care/treatment, 24, 33, 57, 59, 60–63, 164–70; to feminist modes of productive power, 216; to heteronormativity, 191–92, 209, 239; to information, 147; to language, xvi–xvii, 3, 8–9, 41, 72, 97, 114, 162–63, 181, 204, 221, 223

accommodation, female, xvi, 25–26, 178–79

advice: medical, 9, 10, 49, 57, 73–74, 91–92, 98, 130, 168, 229; erroneous, 79, 99–100, 142–45, 146–47, 163

affect. *See* emotions/affect

Aglioti, Salvatore, 201

agnosia, 29, 31, 33, 179–87

Ahmed, Sara, 26

algometry, 65

alienation. *See* genital alienation

"alts" (alternative selves), 213–14

ambivalence. *See* emotions/affect

American Journal of Obstetrics and Gynecology (AJOG), 82, 83, 88

amplification (of vulvar pain and discourse), 25, 29–31, 34, 74, 121, 249, 253

anatomy/anatomies: and body maps, 11, 14, 85–86, 211, 218; *dys*-appearances of, 182–83; erasure of, 19–20, 120–21, 134–35, 157–58; figure, 39; ignorance/confusion about, 20, 21, 24, 29, 31, 41, 104, 115, 153; incomplete nature of, 162; integrity and well-being of, 50–52, 92, 143–51;

INDEX

anatomy (*continued*)
working knowledge of, 221–22, 224; vulvar/genital, xiv, 4, 8, 11, 14, 16, 44–46, 92, 247, 256, 265n4. *See also* bodies, imaginaries, morphologies

Anharrad (patient). *See* participants

anesthesia, topical, 38, 42, 44, 48, 66, 69, 120, 164, 168–69, 200, 257. *See also* lidocaine

anthropology: cultural, xv–xvi, 11, 27, 31–32, 116, 120, 122–23, 129, 168, 241–42, 247, 259n1; feminist, 3–4, 10, 129, 168, 186, 215–19; forensic, 65; medical, 4, 11, 70, 116, 168, 216, 241–44, 247; of the body, 11, 122–23, 134, 156, 215–18, 256

antidepressants, 42–43, 49, 50, 56, 91, 262n4, 263n7, 264n17. *See also* SS(N)RIs

Aristotle, 20–21

assault, sexual. *See* abuse, sexual, and unwanted genital experience

autoimmune conditions, 50–52, 78, 140–42, 265n1, 268n10. *See also* immune system

Bartholin's gland/cyst, 91–92, 97–98, 266n8

Baartman, Saartjie, 21

Beasley, Chris, 6, 191, 212

Bergeron, Sophie, 123

Berlucchi, Giovanni, 201

bioculture/biocultural, 4, 8, 62–63, 79–80, 135, 193, 224, 247, 249, 256

biofeedback: as bodily information, 108–09, 163; as treatment modality, 49, 104, 152, 163, 201, 267n12; author's experience with, 81, 108–14; definition and explanation of, 108–11, 163. *See also* physical therapy

biopsy, vulvar, 44, 53, 54, 119, 142, 148–50, 175, 185, 237

Birth of the Clinic, The (Michel Foucault), 154–55, 241

Blackman, Lisa, 30

blame, self, 93–96, 122–24, 133, 167–69. *See also* emotions/affect

body/bodies: as accumulated, 79–81, 85–86, 87, 106, 115–16, 135, 254; as agent of refusal, 15, 17, 136, 234, 238, 246; as discursive/analytical site, 3, 13, 16, 27, 29, 32, 89, 115, 120–21, 130, 150, 170, 184, 202–03, 239–40; comportment, 25–26, 69, 77, 80–81, 85, 94–95, 109, 116, 136, 139, 208; critical theories of, 11, 67, 156–58, 174–75, 191, 206–09; (dis)identification with, 8–9, 10–11, 23, 28, 29, 106, 203–04; distributed, 14, 85, 202–03, 216–18; genital, 7, 10, 12, 14, 16, 24–28, 80, 85, 88, 99, 111–12, 135, 139, 143, 170, 182, 200, 212–19, 236; hexis, 27–28, 97, 101, 108; historico-cultural dimensions of, 19–21, 31–34, 79–80, 89, 97, 107–08, 139, 155, 239; (hyper)vigilance, 27–28, 257; ignorance/knowledge, 9, 19, 22, 53–54, 57, 69, 104, 153, 168, 181; (un)incorporation of, 9–10, 87–88, 96, 162, 171, 201, 207, 220–23, 257; integrity, 30, 87, 162, 177, 185–87, 199, 203, 256; intragenital dynamics of, 4, 13; ownership, 84, 120, 201, 203–04; potential/capacity for action, 46, 86, 97–98, 129–30, 137, 165, 179–81, 186, 196, 208, 213–14; racialized, 60–63, 65, 79–80, 115, 127, 245–47; sexual vs. genital, 5–6, 14, 24, 30, 71, 89, 96, 171–72, 197, 255; three bodies (Scheper-Hughes and Lock), 11, 30, 190, 193, 216–17. *See also* anatomy/anatomies, body image, embodiment, female sexual/genital

body, imaginaries, mind-body, morphologies, phenomenology
body image, 11, 13–14, 34, 84–86, 179–80, 186; online and offline, 85–86, 97; emerging, 162, 196, 202, 204–05, 208–10, 212, 215; female, 177, 180, 182–83, 185, malleability of, 14, 85, 180, 201–03, 215–16; multiple dimensions of, 180, 212–13; redistributed, 218
Boellstorff, Tom, 213–14
Botox, 254, 273*n*3
Bourdieu, Pierre, 27–28
Braidotti, Rosi, 175, 208–09
Brigette (patient). *See* participants
Brown, Laura, 25–28, 177–78, 180
Brown, Lisa, 17
Butler, Judith, 3, 79, 185

Cacchioni, Thea, 6, 272*n*5
Carpenter, Laura, 268*n*8
Cathy (physical therapist). *See* participants
chronic pelvic pain (CPP). *See* pelvic pain, chronic
Cixous, Hélène, 138–39
Clair (patient). *See* participants
Clarke, Adele, 19
"clean space," xvi, 17, 24, 29, 89, 171, 208
Clément, Catherine, 138–39
clinic, vulvar. *See* Vulvar Health Clinic (VHC)
clitoris, 13, 16, 19, 29, 34, 52–53, 96–97, 119, 140, 148–49, 256; figure, 39
clothing: as aggravating factor, 3, 47, 78, 116; loose-fitting, 78, 172
coitus, 5, 16–17, 41, 67, 101–02, 104, 110–11, 144, 177, 183, 235; coital imperative, 14–15, 41, 65–66, 177, 188–89, 191
Colebrook, Claire, 217
Colleen (patient). *See* participants

comorbid conditions, 193, 252, 264*n*18
comportment, bodily. *See* body/bodies
conferences, medical, 19, 33, 120, 151–59, 244, 249, 251–54, 269–70*n*21, 270*n*3
contraception, 10, 24, 31, 71, 87, 183–84, 243, 265*n*22
corporeality, 16, 28, 32, 79–81, 114, 116, 175, 196–97, 203. *See also* body/bodies, imaginaries, morphologies
craniosacral therapy (CST), 104–05, 129–30, 244, 267*n*12, 267–68*n*6
culture: as discourse, 12, 184, 239–40; as milieu for genital pain, xv–xvi, 3, 9, 10, 14, 61, 82, 96, 108, 115, 120, 126, 128, 138, 145, 152, 155, 169–70, 186, 240–41, 244, 248, 253, 256; as site of intervention, 13, 17, 34, 85, 113, 136, 158, 184, 202, 240–42; assemblage, 138, 249; attitudes, cues, and stereotypes, 61–62, 89, 150, 167–68, 191, 204; contact, and ideas about race, 64, 135, 157; encounters, 5, 28, 74; norms and institutions, 3, 11, 81, 157, 180, 184, 190, 239, 256; popular, 13, 19–22, 168, 180, 186, 238–39, 249; rape, 25–26, 178; work done by vulvas/vulvar pain, 16, 25, 33, 64–65, 82, 100, 120, 216, 239. *See also* bioculture
cunt, xiv, 138, 170–71, 235
Cuvier, Georges, 21
Cvetkovich, Ann, 126, 128

Daphne (patient). *See* participants
Davis, Simone Weil, 22–23
Davis, Wendy, 19
deep-tissue discourses, 81, 97, 101–14, 120, 136, 147, 150
Deirdre (patient). *See* participants
De Preester, Helen, 86, 203

dermatology/dermatologists, 50, 141–43, 150, 152, 263n5
desire, sexual, 5–6, 15–17, 50–51, 66, 127, 160, 164, 171, 173–75, 185–93, 210, 226, 232–33, 246
diagnosis, delays in, 3, 5, 47, 52, 73, 99, 165, 243, 255; insurance codes in, 149–50, 165, 206, 236
dialoguing, as form of physical therapy, 129–32, 163–64, 196, 201
diaries, symptom, 90–91
dietary modifications, 49, 263n8
diffraction, 157
dilators, vaginal, 70, 205–06; as vulvar pain actant, 206, 249; image, 72; use in treatment, 104, 107, 110, 115–16, 205–06, 144–45, 163, 186, 211–12, 225–27, 230, 269n15
disavowal, 11, 20–21, 30, 89, 94–96, 130, 150, 157–58, 164, 179, 202
dis-ease: and physiological disease, 116, 153, 230, 253; as feminist critique, 4; as cultural disability, 3, 34, 97–98, 114, 171; as deleterious cultural discourse, 26, 128, 136, 139, 168, 186, 216, 230, 236, 245; as embodied/lived experience, 19–20, 23–24, 25, 28, 112, 130, 207, 218, 243; as term, xv–xvi, 4; definition, 17–24. *See also* genital dis-ease, vulvar dis-ease
disgust, 8, 10, 21, 82–84, 88–89
"down there," 90–91, 97, 98, 104–06, 114–15, 182, 222, 261n14, 266n5
drape sheet, 70, 87, 112, 266n5
dys-appearance, 116, 181–84
dysesthetic vulvodynia (DV), 46–50, 224; definition, 36; demographics/patient characteristics, 46–48, 127–28, 224; quality of pain, 47–48; treatment of, 48–50. *See also* vulvar pain, vulvar vestibulitis syndrome, vulvodynia

Duden, Barbara, 17, 31–32, 96–97, 114, 139
dyspareunia (pain with sex), 24, 56, 167, 176, 217–18

Einstein, Gillian, 46, 244–45, 256, 260n5
embodiment, embodied experience: as theory, 31, 89, 155, 213; as opportunity/predisposition, 188, 198, 200–01, 220; as racialized, 61–62, 80, 115, 135; as research method, 31–32, 111–14, 240–41; of absence and cultural disparagement, 25–26, 29, 114, 136–38, 160, 162, 170, 184, 239–40; of difference, 64, 191, 213, 216–17, 245–47; of gender, 80, 180; of genitalia, 2–3, 8, 29; of knowledge/epistemologies, 40–41, 155, 246; of pain, 32, 124–25, 132–36, 153–54, 270n1; of trauma, 124–37, 160; techniques, 85, 201. *See also* body/bodies, imaginaries
emotions/affect: affect, 8–9, 10, 14, 21, 26–27, 51, 93–96, 107–08, 113–14, 119, 123, 127–28, 131–37, 153, 155, 166, 178, 208–11, 218, 220–21, 230–31, 234, 242; ambivalence, 15–17, 33, 49, 66, 89, 127, 171, 179–80, 222, 225–30, 241, 246; and neurology/nervous system, 66, 108–09, 137–39, 245–46; anger, 10, 16–17, 82, 86, 88, 94, 224–25; as factor in treatment, 50–51, 68–69, 70–71, 73–74, 93–95, 122–24, 162–63; betrayal, 164–65, 169; depression, 42–43, 56, 64, 73, 130, 187, 264n18; despair, 5, 27, 48, 70, 79, 94–95, 232, 240; embarrassment, 112–13, 146, 149, 183, 192, 210, 222; fear, 28, 41, 91, 110, 127–30, 134, 137–38, 176–77, 210; general discomfort, 84, 92, 106–07, 112–13, 146, 170;

INDEX   315

grief, 73, 94–95; growth and generosity, 135, 195, 219–20; guilt, 130–34, 210, 227; sadness and crying, 93–94, 119, 131–32, 141, 176–77, 225; suicidality, 79, 132. *See also* blame, self, and shame
enteric nervous system, 137, 256
erasure/erasures, vulvar, 9, 29, 96, 100, 115, 184; discursive, 89, 93–94, 120–21, 150–51, 158, 170, 181, 184; material, 100, 115, 151, 170, 177–78, 204, 256
eruptions, 33, 67, 120, 136–39, 160, 184, 214
ethnography, ethnographic research: 5, 11–15, 32–33, 37, 67, 80, 93, 139, 150, 153, 186–87, 199–200, 220, 236–39, 241–47. *See also* anthropology, fieldwork, methods
excess, 4, 27, 29, 32, 96–97, 138–39, 257

Facebook, viii, 20
Fausto-Sterling, Anne, 135
female genital/sexual body: as active, 83–86, 108, 131, 177, 220; as lack, 11, 13, 72, 89, 96, 100, 141, 183, 185; devalued/disparaged, 20–22, 25, 27–28, 34, 85–89, 93, 96, 100, 105, 162, 183, 185, 253; disciplined, 13, 23, 29, 89, 99–100, 174, 206; dysfunction of, 27, 185, 233–34; feminist analyses of, 4, 10–11, 16, 32, 84, 89, 116–17, 131, 223–24; fragmented, 150, 162, 171–74, 177; inscrutability of, 99–100, 140, 206–07, 227; intragenital dynamics of, 4, 13; undisclosed nature of, 24, 93; value of, 4, 11, 16, 117, 143, 172, 245. *See also* body/bodies, embodiment
femininity: as mode of bodily comportment, 136; and racialized whiteness, 63–67
feminist/feminism, xv, 27, 63, 80, 86–88, 98, 113, 124–25, 129, 131, 136, 139, 144, 178, 180, 186–87, 190–91, 202, 206, 208, 224, 246–47, 261*n17*; anthropology and, 129, 215–19, 242; corporeal, 16–17, 21, 32–33, 80, 124, 179–80, 187, 208, 215–19, 233, 257; health movement/health care, 9–10, 81–83, 86, 124–25, 168, 243, 260*n6*; politics, 11, 13, 83, 103, 116, 124, 186, 216–17, 243, 246, 260*n6*; postfeminism, 66, 238, 246; science studies, 46, 202, 233–34, 256; second wave/waves of, 9, 246, 260–61*n7*; theory and analysis, 4, 7–8, 32, 63–67, 98, 131, 162, 190, 224, 246–47
fibromyalgia, 51, 56, 78, 252, 265*n1*
Fixing Sex (Katrina Karkazis), 12
Frances (patient). *See* participants
field notes, 27, 81, 98, 120, 133, 147, 152, 154, 169, 176, 195, 212, 219–20, 227, 237, 238
fieldwork, 5, 19–20, 32, 35–36, 45, 80–81, 103, 114–15, 122, 124, 127, 153, 171, 183, 205, 240–43, 259*n2*, 264*n20*, 264*n21*; participant-observation in, 108, 199–200
Foster, David, 152–59, 243
Foucault, Michel, 154–55, 199, 205–08, 223–24, 240–41
Freud, Sigmund, xiv, 8, 30, 33, 101, 124, 128, 131, 136–38, 158–60, 199, 216, 260*n6*
Frueh, Joanna, 88, 235

gender affirmation treatment, 13
gender: and sexuality studies, *x*, 12, 46, 83, 168, 174, 180, 202, 269*n14*; as aspect of vulvar pain, 8, 67, 135, 217; asymmetrical norms and dynamics related to, 7, 12–13, 33, 36, 61–62, 81, 99,

316 INDEX

gender (*continued*)
124, 134–35, 180, 215, 217, 255; embodied nature of, 80, 115, 136, 213–14, 217, 264n18, 265n1; identity, as structured by heteronormativity, 6–7, 12, 14, 33, 83, 177, 211–13; intersections: with race, 55–67, 80, 135, 217, 247; with religion: 190, 215

gender/sex binary, 12, 14, 30, 174, 240

generalized vulvar dysesthesia, 2, 36, 46. *See also* dysesthetic vulvodynia, vulvar pain, vulvar vestibulitis syndrome, vulvodynia.

genetics/genetic predisposition, 43, 58–60, 252; biological markers, 58–59, 252; (Caucasian) phenotype, 57–66, 153, 252

genital/genitals, 20, 22, 28, 96, 112, 114, 128, 135, 141, 143, 145–46, 151, 158, 162, 169, 179–81, 184–85, 190, 196, 201, 204, 216–17, 249; alienation from, 4, 18, 29–31, 38, 41, 81, 136, 152, 154–55, 160, 169, 170–71, 177, 183–85; baggage, 32, 81, 95, 102, 116–17; cutting, 45–46, 238–39, 256, 262n18, 268n11; definition of, 12; difficulty talking about, 18–19, 41, 83–84, 86, 188, 122, 224; dis-ease, xv, 3–4, 8, 17–24, 38, 39, 97–98, 112, 116, 128, 137–38, 168, 171, 218, 243, 271; diversity, 12–14, 46, 175, 180, 218, 248, 260n5; integrity, 27, 85, 87, 92, 147, 116, 166, 177–78; lived experience of, 8, 90; missing, 11, 16, 19, 29, 33, 89, 147–48, 171, 180–81, 185, 224, 248–49; normativity, 150; painful contact with, 27, 41, 48, 153, 183, 195; shame surrounding, 8, 71, 87, 91, 114–15, 167, 179, 191, 221, 248; silence about, 14, 18. *See also* labia

genitality, 11–14

"getting better," 33–34, 122, 161–70, 255

Gia (medical assistant). *See* participants

Goetsch, Martha, 43–48, 56–58, 263n6, 266n8, 267n3

gold standard, clinical, 156, 158

Grosz, Elizabeth, 85–86, 136, 175, 179–80, 190

gynecology/gynecological medicine, 3, 9, 23, 33, 56–60, 63, 67–69, 71–72, 82–83, 86–89, 91–92, 101–03, 136, 152, 156, 184, 221, 240, 252, 255, 259n1, 259n2, 267n2, 270n6; as heteronormative, 11, 15–16, 19, 24, 72, 75, 182–83, 235; as patriarchal, 11, 21–23, 99, 184, 234, 266n7; author's relationship to/history with, 3, 9–11, 52, 71–72, 81, 84–89, 112, 113 116, 168–69, 183–84, 186, 259n1, n2, 267n3; residents, 35, 60–63, 224; teaching associates (GTAs), 82–84, 265n4

habitus, bodily habits, 19–20, 31, 84–85, 89, 92, 95–97, 136, 152, 170–71, 187, 208. *See also* body/bodies, hexis

Haefner, Hope, 52

Haggerty, Catherine et al., 62

Hanna (physical therapist). *See* participants

Haraway, Donna, 64, 156–57

Harlow, Bernard, 55, 58–60, 127–30, 134, 170, 242–43, 248, 252, 267n5

Harvard School of Public Health (research studies), 1, 3, 58–59, 107–08, 127–29, 170, 224, 242–43, 247–48, 252, 267n5

heteronormativity/heterosexuality, 13–18, 32–33, 55, 81, 101, 136, 140, 174–78, 185–88, 191–92,

211, 215–16, 218–19, 224, 233–35; ambivalence toward, 15–16, 66, 225–30, 246; as social structure, 5–7, 11, 18, 182, 207; coital imperative of, 14–15, 177; critical analyses of, 14–17, 58, 175, 246–47; disruption of, 4, 6–7, 22–30, 72, 185, 190, 200, 211–13, 254; queering of, 6–7, 175, 206, 239; routine violence of, 17, 63, 88, 134–35, 160, 178, 241; scripts, 4, 13, 150, 174, 177, 182, 233. *See also* penetration
Hess, Amanda, 26
hexis, bodily, xvi, 27–28, 31, 97, 101, 108, 113
History of Sexuality, The (Michel Foucault), 206, 240
holding patterns, 81, 107–08, 111–17, 122, 163, 201
home and family, 10, 128–29, 132–34, 161, 166–69, 174, 192, 222–23; as amorphous entity, 168; as code for sexual intercourse, 168–69, 189; as site for intervention, 161, 168
Horn, David, 64–65, 246
House of Representatives: Michigan, 17; Texas, 19; United States, 26
hypochondria, 49–50, 165, 176, 183, 238
hysteria, 63–67, 137, 159, 245–46

imaginaries: alternative, 3–4, 14–16, 73, 97–98, 115–16, 121–22, 136, 200, 205, 207, 212–13; bodily, xiv, 4, 46, 85, 116–17, 121, 139, 145, 196, 207, 209, 213; clinical/medical, 61–63, 96–97, 159, 245; genital/vulvar, 14–16, 96–97, 145, 198, 205, 212, 217, 218–19; (hetero)sexual, 6, 15, 33, 66, 81, 97, 174–75, 185–86, 191–92, 196, 207–09, 216, 234–35, 246; racialized, 20, 61–63, 245
immune system, 43, 49, 51–52, 141–42, 154, 66, 91–92; modulators, 50, 93, 141–46, 268n12
incompetence, structural, 151
infibulation, 140, 268n10. *See also* genital cutting
inflammation: as eruptive, 139; as nature of vulvar pain, 38, 42–44, 50–54, 59, 66, 135, 140, 144, 148–49, 262n3; in Bartholin's cyst infection, 91–92
insurance, health, 44, 49, 69, 73–74, 101–02, 141, 148–50, 165–66, 193, 233, 236, 239, 249, 270n3; diagnostic codes, 165, 206, 236
integrative medicine, 68, 122, 129, 173
interdisciplinarity, 241–45, 252
intersectional mode of analysis, 21, 60, 62–63, 65, 148, 245–47, 256
intersexuality, 12, 14, 45–46, 175, 180
introitus, vaginal, 38, 44, 140; figure, 39
invisibility, xiv, 18–19, 29, 74, 90, 142, 150–51, 188, 247. *See also* absence
Irigaray, Luce, 16, 18, 171, 174, 182, 187, 196, 224, 247
Isabelle (patient). *See* participants

Jackson, Stevi, 6, 187, 211, 213
Jane (nurse). *See* participants
Jessica (patient). *See* participants
Jill (sex therapist). *See* participants
Joan (patient). *See* participants
JoJo (patient). *See* participants
jokes, dirty, 8, 11, 21, 26–28
jouissance, 173, 270n5
Journal of the American Medical Women's Association, 1
Judy (patient). *See* participants

Kaler, Amy, 7, 177
Kapsalis, Terry, 81–84, 88–89, 265n22, 266n7
Karkazis, Katrina, 12–13, 46, 260n5

Kempner, Joanna, 7, 40, 61–62, 135, 264*n*18, 265*n*2, 272*n*1, 273*n*4

Lacan, Jacques, 85, 173, 270*n*5
labia, 4, 20, 29, 51–54, 96–97, 119, 139–44, 148–50, 178, 181, 218, 225, 235; figure, 39. *See also* genitals
labiaplasty, cosmetic, 22–24, 138, 164, 224, 238–39, 256, 262*n*18; complications from, 23; feminist analyses of, 224, 238–39, 243, 261–62*n*17. *See also* genital cutting, surgery
language: difficulty with/"talking about it," 3, 5, 8–9, 17–18, 21–22, 24, 41, 49–50, 57, 72, 73, 79–82, 93–94, 98–100, 106–07, 110–11, 114–16, 119, 126–27, 181, 187–88, 195, 204, 215, 217–18, 220, 221, 224–30, 254, 256, 278*n*8; naming, xv, 1, 3, 92, 97, 119, 122, 131–32, 197–98, 241, 254
laser, pulse-dye, 44, 48–49, 164, 209, 263*n*5. *See also* treatment
Leder, Drew, 182
Lerner, Harriet, 8–9
Libby (patient). *See* participants
lichens, 37–38, 50–55, 163–64, 224; planus, 37, 51–54, 114, 139–48, 185, 222, 225, 269*n*16; sclerosus, 34, 37, 50–54, 97, 140, 148; simplex chronicus, 37, 50, 54–55
lidocaine, 38, 42–43, 48, 66, 91, 94, 104, 142, 164, 168–69, 209, 230. *See also* anesthesia, topical
Lisa (physical therapist). *See* participants
Lock, Margaret, 216
Lori (patient). *See* participants
Louise (patient). *See* participants

map, body. *See* body image
marriages/marital disruption, 25, 32, 73, 79, 105, 121, 133–34, 136, 163, 183, 189–90, 210, 225, 229–30, 239, 241, 248. *See also* partners
Mary Hudson (patient). *See* participants
matter, materiality, *xiv*, 4, 8, 11, 13, 16–18, 32, 67, 72, 77, 80, 87, 99, 115, 120, 141–45, 149–50, 157, 162, 170, 178, 191, 201, 203–06, 209, 214, 217, 234, 240, 246, 252, 256–57
McEwan, Melissa, 26
medical assistants, *ix*, 35, 68–70, 74–75, 89–90
medical history/intake forms, 2, 8–9, 41, 48, 54, 68–69, 74–75, 77–78, 89–90, 98–99, 119–20, 123–27, 164, 183, 193, 254–55
Merleau-Ponty, Maurice, 179, 197–200, 202
methods, research, 5, 19–20, 32, 35–36, 58, 220, 240
migraine, 7, 61, 135, 264*n*18, 265*n*2, 272*n*1, 273*n*4
mind-body, 30, 122–26, 159
misogyny, 21, 83; online, 26; medical, 146–47
Mitchell, S. Weir, 64
Moore, Lisa Jean, 19
morphologies, 3, 32, 174, 180–81, 198–205, 207, 215. *See also* imaginaries
Muscio, Inga, 170–71
Mya (patient). *See* participants
myalgia, pelvic floor, 101–04, 159, 165, 193, 216, 236. *See also* physical therapy, vaginismus
My New Pink Button, 20

National Institute of Child Health and Human Development (NICHD), 251–55
National Institute of Dental and Craniofacial Research (NIDCR), 251

National Institutes of Health (NIH): 2004 conference, 37, 151–60; 2006 state of the art report, 37; 2011 conference and report, 251–56; influence on vulvar pain funding and research, 151–55, 224, 243

National Vulvodynia Association (NVA), 50, 57–58, 102–03, 151, 155, 206, 224, 243, 263n9, 267n4

neurobiology/neuroanatomy, 46, 137–39, 154–62, 201–4, 255–56, 260n5

neuromodulators/neuroleptic agents, 42, 50, 56, 66, 92–93; side effects of, 49, 92–93, 263n7

neurology/nervous system, 33, 42, 54–56, 63–66, 64–67, 108–09, 129, 137, 154–62, 213–14, 245–46, 256

neuroma, 91–93

neuropsychology/neuroscience, 10–11, 29, 33–34, 63–66, 85, 179, 198, 201–04, 255–56, 273n4

Nikki (patient). *See* participants

normality, challenges to and struggles with, 6, 15–17, 63, 73, 163, 175, 184, 191–92, 200, 210, 216, 220, 231, 239

nursing/nurse practitioner, 9, 53, 148, 237–38; author as, 9–10, 81–89, 113, 168, 183, 186, 259n1, n2; vulvar, 70, 73–74

obligation, xvi, 11, 30–31, 33, 107, 136, 145–47, 149, 155, 184, 248–49, 256, 270n1

"off-label" medications and treatments, 44, 74, 141–47, 269n13. *See also* pharmacists/pharmaceuticals

Onion, The, 21–22

Orange is the New Black, 20

orgasm/anorgasmia, 19, 23, 29, 34, 45–46, 49, 53, 137, 171–72, 231–34, 246, 272n3

pain, not specific to genitalia: as alienating force, 29–30, 131–34, 135, 177, 194; as catalyst for change, 6–7, 116, 181; as bodily index, 17, 130, 137–38; perception and pharmacological management of, 42–43; phantom, 29, 179, 184, 203; race and, 64–67, 127, 217–18, 245–47; clinical research, 153–56, 158, 251–52, 254–56

participants: Anharrad, 146, 225–27, 229–30, 234, 254, 272n3; Ashley, 181–82, 222, 228–29; Brigette, 184–85, 187–88, 219–20, 222; Cathy (physical therapist), 93, 103, 105, 109–14, 161–63, 166, 169, 201–02, 204–05, 210–15, 224, 267n12, 268n6, 270n2; Clair, 105–08, 169, 191–92, 219, 230–33, 240, 254, 272n4; Colleen, 175–78, 185, 254, 271n6; Daphne, 54, 161, 195–96, 200–01, 208; Deirdre, 98–99; Frances, 89–97, 145, 266n9; Gia (medical assistant), 69, 74, 90, 96; Hanna (physical therapist), 93, 108–10, 161, 196–98, 201, 204, 244; Isabelle, 57, 98–99, 188–92, 195, 220–22; Jane (nurse), 70, 73–74; Jessica, 161, 210, 225; Jill (sex therapist), 71, 73, 123, 205, 215; Joan, 97, 147–50, 204, 235; JoJo, 98–99, 239; 268n11; Joy (physical therapist), 161, 204; Judy, 139–50, 178, 204, 254, Libby, 161, 201–02, 205, 210–18, 224, 238, 271n2; Lisa (physical therapist), 121–22, 129–34, 197, 204, 267n12, 267n1; Lori, 23; Louise, 77–79, 90; Mary Hudson, 53, 140, 148, 150, 204; Mya, 166–70, 221, 223, 227, 230; Nikki,

participants (*continued*)
121–23, 128–36, 170, 197–98, 240, 248, 254, 267*n1*; Rosemary, 183; Scout, 171–74; Sharon, 4–7, 184–86, 225–34; Susan, 231; Tina, 222–23

partners/husbands, 5–7, 15–18, 22, 41, 49–51, 63, 100, 167–69, 182, 185, 195, 201–02, 209–10, 219–20, 224–30, 246, 271*n8*, 272*n3*. *See also* marriages/marital disruption

patient vs. woman role, 23–24

pelvic exams, ix, 1, 81–88, 101–14; models for (GTAs), 82–84, 265*n4*

pelvic floor, 28, 43, 48–49, 81, 96, 100–07, 110–12, 138, 162–65, 193, 196–99, 210–15, 214–18; muscles of: figure, 102; levator ani, 164; pubococcygeal, 102, 164

pelvic pain, chronic (CPP): definition, 62; and race, 61–63, 217–18, 245–47

penetration, vaginal-penile: ambivalence toward, 16–17, 66–67, 225–30; as cultural narrative/orientation, 4–6, 12, 15–16, 19, 24, 27, 49, 51, 143–44, 173, 177, 185, 234; as pain or violence, 26–27, 119–20, 124–25, 128, 163, 224; as part of treatment, 42, 106–07, 144, 163, 168, 211–13, 226, 230; as "real sex," 5, 19, 24, 41, 143, 177, 185, 189, 195, 211, 228, 239; as sacred, 188–90; as ineffective route to orgasm, 246, 272; inability to tolerate, 3, 15, 32, 43, 47, 62, 96, 103–04, 140, 143, 153–54, 167, 176, 224; penetrative triumvirate (penis, tampon, speculum), 1, 41, 101, 266 *n10*; refusal of, 124, 136–37, 186–87, 210, 235; vulva's role in, 182, 211–13, 234; with tampons, 1, 41, 101, 167. *See also* coitus,

coital imperative, heteronormativity/heterosexuality, refusal

phallus/phallocentrism, 4–6, 16, 21, 138, 173–75, 188, 204, 209–11, 219, 233

phantom pain and sensation, 29, 100, 179, 184, 203

pharmacists/pharmaceuticals, 7, 74, 142–44, 146–47, 150, 188, 225, 233, 249, 269*n18*, 272*n1*; compounding pharmacies, 269*n16*

phenomenology, 163, 179, 198, 200–03

physical therapists (PTs): as informants, 35, 111–14; as part of health care team, ix, 74, 120–21, 205; local providers, 93, 109, 121, 161, 169–70; relative expertise of, 103–05, 107–09, 121, 123, 129–30, 137, 158–59, 201, 215, 236, 244, 248, 253, 267–68*n6*. *See also* physical therapy

physical therapy (PT): as access to alternative imaginaries, 16, 33, 97–98, 173–74, 198, 121–22, 196, 201–05, 209–15; as intensely personal, 15, 33,100, 106–07, 121, 162–63, 191–92, 196; author's experience with, 111–14; barriers to, 33, 106–08, 166, 169–70; biofeedback and, 104, 108–11, 201; indications for, 43, 45, 48, 81, 101–11, 163–64, 271*n2*; insurance coverage of, 102–03, 165, 236; mechanisms in, 100–11, 159, 163–64, 201; preparation for, 42; referrals to, 92–93, 101, 121, 164–66, 169, 236; relative efficacy of, 34, 103, 108, 123, 159, 161–62; transcript of session, 202, 205, 211–12, 214–15; types of, 103–05, 137, 163. *See also* biofeedback, craniosacral therapy (CST), dialoguing, holding patterns, physical therapists

Pi-Chacán, xiii–xv. *See also* vulvar sculpture
Planned Parenthood, 9, 19
Pollock, Anne, 61–62, 268*n*9
pornography, 22–23, 29, 178, 192, 249
postfeminism. *See* feminist/feminism
Potts, Annie, 174–75, 190–91
power: as biopower, 206–07; as generative/productive, 199, 205–06, 207. *See also* Foucault, Michel
prevalence. *See* vulvar pain prevalence
psychoanalysis, 124–25, 130, 137–38, 173, 199. *See also* Freud, Sigmund
Psychosomatic (Elizabeth Wilson), 137, 195
Public Privates (Terry Kapsalis), 82

q-tip/swab test, 38–41, 44, 48, 56, 165, 263*n*6; figure, 40
queer theory/queerness, 6, 16, 46, 47, 126, 173–75, 186–91, 200–01, 206, 211, 239, 248

race, 20–21, 32, 41, 47; as bodily accumulation, 79–80, 135; chronic pelvic pain and, 61–63, 217–18, 245–47; critical theories of, 32, 63–67, 156–57; hysteria and, 63–67; vulvar pain and, 56–63, 127, 217–8, 245–47; whiteness as, 32, 55–67, 217–19, 245–47
rape culture, 25–26, 178
refusal: xv; ambivalence toward, 160, 169–71, 210, 225, 246; bodily forms of, 15, 65–66, 136, 179, 234; of heteronormativity and vaginal intercourse, 6, 15–17, 65–66, 136–37, 186–87, 190, 210, 234, 246
religion, 41, 45, 107, 127, 172, 189–90, 192, 215, 245, 248

reproductive capacity/childbearing, 14, 144, 174, 190
"Research Plan on Vulvodynia," (NIH 2012 report), 251–56. *See also* National Institutes of Health (NIH)
Rokudenashiko ("no good girl"), 17
Root, Maria, 95, 178
Rosemary (patient). *See* participants

Sacks, Oliver, 33, 213–14, 220, 235
"safety and support," 127–29, 170, 242, 247–48
Savage, Dan, 54, 264*n*13
Scarry, Elaine, 29
schema, body. *See* body image
Scheper-Hughes, Nancy, 216
Scheurich, Neil, 63, 66
Scout (patient). *See* participants
Segal, Lynne, 173, 186, 190, 270*n*5
sex education, inclusion of vulvar health in, 181, 243, 254–55
sex reassignment surgery. *See* gender affirmation treatment
sex therapy, 45, 73, 101, 123–25, 210–11, 215, 271*n*2
Sex Variant Study, 20
sexual abuse. *See* abuse, sexual
shame, genital/bodily, 8–9, 20–22, 28, 32, 69–71, 82, 88, 91–96, 113–15, 169–71, 180, 191, 249, 257. *See also* emotions/affect
Sharon (patient). *See* participants
Shildrick, Margrit, 244
Showalter, Elaine, 64–65
"shutting down," xvi, 5, 15, 17, 30, 34, 48, 197, 200; as heaviness, 80; as stagnation, 17, 95, 100. *See also* emotions/affect
Smith, Yolanda, 52
Sobchack, Vivian, 182, 200, 203–04
social science/social scientists, 13, 60–62, 152, 242–44, 254–55
Solnit, Rebecca, xv, 146

somatic: locations for feelings/somatization, 130, 160, 247; modes of attention, 136, 198; sociosomatic eruptions, 160
speculums, 41, 84–85, 120, 168, 267n2. *See also* penetration
SS(N)RI's, 42–43, 49, 55, 262n4, 263n7. *See also* antidepressants
Steinem, Gloria, 8
steroids, 49–55, 141–43, 148, 156
Stewart, Elizabeth Gunther, 55, 58–59, 127–30, 134, 170, 242–43, 248, 252, 267n5
surgery 5, 12–14, 22–24, 31, 41–46, 48–49, 54, 101, 121–23, 139–45, 148–49, 164, 176–79, 200, 209, 232–33, 237–39, 244, 261n9, 262n18, 263n6, 266n9, 272n4; complications, 91–96. *See also* genital cutting, labiaplasty, vestibulectomy
Susan (patient). *See* participants

"tail-tucking" maneuver, 27–28, 96, 101. *See also* pelvic floor
tampons, 1, 41, 101, 167, 183. *See also* penetration
Tapper, Melbourne, 61
Teeth (film), 20, 261n14
Things They Carried, The (Tim O'Brien), 94
Tiefer, Leonore, 22
time/temporality, 47, 68–69, 96–97, 131–34, 138–39, 155–56, 164, 198–201, 204–05, 208–10, 240
Tina (patient). *See* participants
transcripts, xvi–xvii, 80, 199. *See also* field notes, fieldwork, methods
transgender, 12, 14, 20, 45–46, 175, 180, 218, 248, 260n5
trauma, 25–26, 101–02, 124–29, 135, 139, 177–78, 215–16; embodied, 27, 48–49, 93, 124–29, 129–30, 135, 156; insidious, 95, 178–79; PTSD, 98, 128; sexual, 124–29. *See also* abuse, sexual, and safety and support
treatment. *See* vulvar pain treatment
Tsakiris, Manos, 86, 203

ultrasound, transvaginal, 26–27
unwanted genital experience: 4, 10, 18, 24–28, 32, 81, 96, 114, 116, 122–23, 126–36, 160, 208, 242, 248. *See also* abuse, sexual
urinary problems, 54, 103, 114, 139–45, 263n8, 268n11

vagina/vaginal: as offensive word, 17, 235; as "open"/penetrable space, xiv–xv, 12, 51–54, 140–45, 150, 182, 226, 233, 259n4; as part of alternative genital imaginary, 202–18; complementarity with penis, 72, 154, 189–90, 239; cosmetic surgeries of, 22; dentata, 261n14; dilators, 71–72, 116, 144, 230; figure, 39; in bodies of friends, 84, 86–87, 112; muscle tightening, 101–05, 110, 114, 162–64; opening of (introitus) 38, 178; orgasms, 19; non-penile penetration, by: ultrasound, 26–27, biofeedback sensor, 110–12, 163, finger, 162–63, 196–97, 266n10; poor understanding of anatomy, 146, 183–84, 221; suppositories, 50–51, 145–47; vs. vulva, xiv–xv, 4, 8–9, 11, 13–14, 19–20, 97, 148–49, 180–81, 221, 261n10. *See also* dilators, vaginal, introitus, penetration, and vaginismus
Vagina Monologues, The (Eve Ensler), 8
vaginismus, 101–03, 106, 124–25, 236. *See also* myalgia, pelvic floor
Valenti, Jessica, 26
Veronica, 129–36, 198
vestibule, vulvar, 36, 38, 41, 104; figure, 39
vestibulectomy, 44–46, 232, 237,

239, 263n6; figure, 45. *See also* surgery
vestibulodynia, 37, 38. *See also* dysesthetic vulvodynia, vulvar vestibulitis syndrome, vulvodynia
Viagra, 146, 233, 269n17
virginity, 41, 107, 167, 195–96, 210, 268n8
Volatile Bodies (Elizabeth Grosz), 179,
vulnerability: female, 10, 28, 34, 119, 128, 135; medical/professional, 99–100, 111, 255; vulvar/genital, 9, 11, 95, 106–07, 119, 184, 238–39, 241
vulva/vulvar: awareness, 22, 57, 82, 97, 100, 116, 122, 125, 140, 148, 151–53, 181–82, 196, 201–02, 220–22, 224, 248; -based sexuality, 7, 16, 71, 184–93; biological variations of, 12–14, 20, 45–46, 80, 180, 191, 216–18, 233–34, 248, 256–57, 260n5, 262n18; biopsy, 44–45, 53, 54, 119, 142, 148–50, 175, 185, 237; contour, change of, 51–54, 97, 119–20, 140–48, 148–51; culturally charged, 33, 82, 124, 155, 257; dis-ease, xv, 20–21, 114, 116, 168, 230, 253; disinvestment and inconsequence, 4, 9, 11, 52, 79, 89, 91–98, 100, 113, 117, 136, 141–51, 155, 158, 162, 174, 178, 184, 197, 203–04, 218, 249, 255, 266n9; disparagement and distaste toward, xv, 3, 11, 15, 20–21, 26–29, 32, 69, 79–80, 82, 85, 88, 96–97, 113–15, 126, 134–36, 150–51, 162, 170, 183, 218, 239, 253–56; distributed, 14, 202–03, 216–18; erasures, xv, 8, 19, 29, 34, 93–96, 134–35, 151, 167–70, 184; excisions of, 13, 22–24, 27, 44–46, 175, 178, 183–84; experts/expertise, 30, 48–49, 50–51, 57, 60, 82, 125, 129, 144, 159, 205;

imaginaries, 14–16, 71–72, 119, 121–22, 145, 196–98, 207–08, 213, 218; integrity, 26–27 (threats to), 50–51, 52, 92, 177; malignancy/cancer of, 51, 175, 183, 270–71n6; missing, 11, 16, 19, 29, 33, 147–48, 171–73, 180–81, 185, 224, 248–49; neurobiology of, 46, 245, 256; non-compliance, 6, 15–16, 33, 183, 207–08; perplexing nature of, 48, 52, 91, 127; popular culture/representations of, 13, 20–22, 89, 228; queered, 16–17, 21, 126; recollection and recuperation of, 4, 16, 135, 202, 220; reluctance toward/ignorance of, 8–9, 14, 21, 29, 41, 49–50, 52, 152–54, 173, 222–23; sculpture (Pi Chacán), xiii–xv; scholarship, lack of, 13, 82, 90, 124–25, 173; self-examination of, 116, 149, 152; shame, 8–9, 20, 22, 69, 92, 97–98, 106–07, 114–16, 169, 171, 180; skin care measures, 45, 50, 177, 231; vs. vagina, 4, 8–9, 13, 19–20, 216, 221, 235; vulnerability, 9, 99–100, 106–07
vulvar dis-ease, xv, 3–4, 17–24, 38, 39, 114, 120, 128, 168, 171, 218, 230, 271; definition, 8. *See also* dis-ease, genital dis-ease
Vulvar Disease, as formal concept, 8, 34, 241, 249
Vulvar Health Clinic (VHC), 5, 30, 35–38, 40–41, 43–44, 46–48, 50, 57, 60–63, 67–76, 77–78, 96–98, 101, 103, 123–26, 139–41, 145–53, 159, 164–66, 169, 232–33, 235–38, 244, 246–47, 262n1, 264–65n21
vulvar pain, 247–49; affective dimensions of, 42–43, 51, 79, 133–34; as cultural condition, 3, 79–81, 100, 120, 128, 206; as medical specialty, 120–25, 254–55; as refusal of penetration, 103,

vulvar pain (*continued*)
210, 234, 246; as pain (vs. sexual) condition, 252, 255–57; characteristics of, 36, 41, 46–48, 78, 81, 119–21; clinical research about, 56, 58, 82–83, 120–25, 151, 251–56; definition of, 1, 36–37, 46–47; demographic/patient profile, 32, 41–42, 46–47 (DV), 55–59, 103, 104–07, 114, 125–28, 134 (VVS), 152, 247; etiologies (causes of), 2, 43, 48, 55–56, 78–81, 100, 127–30, 134, 156, 159, 199, 243; feminist analyses of, 12, 81, 124–25, 131, 224; neuropathic nature of, 42–43, 63–67, 120; nomenclature, 36–38 (table), 206, 218, 245; "other causes" of, 2, 17–18, 18–31, 28, 79, 123, 134, 167, 252; paradoxical nature of, 34, 99, 135, 149; physical exam for, 38–41, 176 (VVS); physiological reality of, 2, 31, 35, 43, 78, 81, 115–16, 120, 122–23, 130, 134, 153–56, 165, 176, 243–44, 247; population–based research about, 1–2, 56–60, 243; prevalence of, 2, 56, 57–59; as psychosomatic condition, 55–57, 98, 101–02, 124–26, 156, 237–38; race and, 41, 55–63, 217–19, 245–47; screening tools for, 139, 148, 153, 248; symptoms of, 1, 5, 38, 78, 91, 94, 96, 98, 119–20, 124, 142; treatments for, 42–46 48–49, 92–94, 98–100, 109–10, 123–24, 153, 163–64, 200, 209, 225, 230. *See also* dysesthetic vulvodynia and vulvar vestibulitis syndrome

vulvar vestibulitis syndrome (VVS): definition and characteristics, 36–37, 38–46, 163; diagnostic test for, 38–41, 153; patient profile, 41–42; treatment of, 42–46, 164. *See also* vulvar pain

vulvodynia. *See* dysesthetic vulvodynia and vulvar pain

Vulvodynia: A Chronic Pain Condition—Setting a Research Agenda (2011 NIH sponsored conference), 251. *See also* National Institutes of Health

Vulvodynia and Sexual Pain Disorders in Women (2004 NIH sponsored conference), 151–60. *See also* National Institutes of Health

VVS. *See* vulvar pain and vulvar vestibulitis syndrome

"war on women" in United States House of Representatives, 26
Williams, Raymond, 26
Weismantel, Mary, 79–80
whiteness. *See* race
Wilson, Elizabeth, 30, 33, 137–38, 174, 185, 195, 234, 248–49, 256

Young, Iris Marion, 79, 136
Young, Katharine, 88, 155